NMR

Basic Principles and Progress
Grundlagen und Fortschritte

13

Editors: P. Diehl E. Fluck R. Kosfeld

Editorial Board: S. Forsén S. Fujiwara
R. K. Harris C. L. Khetrapal
T. E. Lippmaa G. J. Martin A. Pines
F. H. A. Rummens B. L. Shapiro

Introductory Essays

Edited by M. M. Pintar

With 48 Figures

Springer-Verlag
Berlin Heidelberg New York 1976

Dr. Milan Mik Pintar

University of Waterloo, Faculty of Science
Department of Physics
Waterloo, Ontario N2L 3G1/Canada

ISBN 3-540-07754-5 Springer-Verlag Berlin Heidelberg New York
ISBN 0-387-07754-5 Springer-Verlag New York Heidelberg Berlin

Library of Congress Cataloging in Publication Data. Main entry under title: Introductory essays. (NMR, basis principles and progress; v. 13) Bibliography: p. Includes index. 1. Nuclear magnetic resonance-Addresses, essays, lectures. 2. Spin-lattice relaxation-Addresses, essays, lectures. I. Pintar, M. M., 1934– II. Series. QC490.N2. vol. 13. [QC762] 538'. 3s. [538'.3]. 76–15663.

This work is subject to copyright. All rights are reserved, whether the whole or part of the material is concerned, specifically those of translation, reprinting, re-use of illustrations, broadcasting, reproduction by photocopying machine or similar means, and storage in data banks. Under § 54 of the German Copyright Law where copies are made for other than private use, a fee is payable to the publisher, the amount of the fee to be determined by agreement with the publisher.

© by Springer-Verlag Berlin · Heidelberg 1976.
Printed in Germany.

The use of registered names, trademarks, etc. in this publication does not imply, even in the absence of a specific statement, that such names are exempt from the relevant protective laws and regulations and therefore free for general use.

Typesetting and printing: Carl Ritter, Wiesbaden. Bookbinding: Brühl, Gießen.

Editorial Guidelines

Since the Series "NMR — Basic Principles and Progress" was founded in 1969 it has dealt primarily with the theoretical and physical aspects of the methods. Today nuclear-magnetic resonancespectroscopy has become one of the principal techniques of the chemist and is finding increasing use in the fields of Biology, Pharmacy, Medicine, and Criminology. The growing significance of applied spectroscopy has earned it a correspondingly important place for the future in this Series. With the aim of achieving a balanced representation of theoretical and practical problems and results, the present Editors have asked several world-renowned scientists in the field of NMR-spectroscopy to join an International Editorial Board.

The international nature of this Board will facilitate closer contact among research groups and authors throughout the world, making it possible to follow comprehensively the developments in pure and applied NMR-spectroscopy. On this basis, the readers of the Series will be assured of up-to-date contributions not only of current significance, but of long-term value as well.

Prof. E. Fluck, Prof. P. Diehl, Prof. R. Kosfeld, 1976

Preface

This set of essays was given as lectures at the 4th Waterloo International Summer School on Nuclear Magnetic Resonance held in June 1975 at the University of Waterloo.

The school was sponsored by the National Research Council of Canada and by the Canadian Association of Physicists. These Contributions are introductory and were not intended to be review papers.

For valuable help, I would like to thank R. S. Hallsworth, D. W. Nicoll, and R. T. Thompson.

M. M. Pintar

Table of Contents

A Guide to Relaxation Theory. By A. G. Redfield 1

Thermodynamics of Spin Systems in Solids. An Elementary Introduction.
By J. Jeener.. 13

Coherent Averaging and Double Resonance in Solids. By J. S. Waugh 23

Macroscopic Dipole Coherence Phenomena. By. E. L. Hahn 31

Nuclear Spins and Non Resonant Electromagnetic Phenomena
By G. J. Bene.. 45

Nuclear Spin Relaxation in Molecular Hydrogen. By F. R. McCourt 55

Longitudinal Nuclear Spin Relaxation Time Measurements in Molecular
Gases. By R. L. Armstrong 71

Spin-Lattice Relaxation in Nematic Liquid Crystals Via the Modulation of
the Intramolecular Dipolar Interactions by Order Fluctuations. By R. Blinc 97

NMR Studies of Molecular Tunnelling. By S. Clough 113

Effect of Molecular Tunnelling on NMR Absorption and Relaxation in Solids.
By M. M. Pintar 125

How to Build a Fourier Transform NMR Spectrometer for Biochemical
Applications. By A. G. Redfield 137

The contributions evolved from the "International Summer School on Nuclear Magnetic Resonance", Department of Physics, University of Waterloo, June 23rd to June 28th, 1975.

List of Editors

Managing Editors

Professor Dr. Peter Diehl, Physikalisches Institut der Universität Basel, Klingelbergstr. 82, CH-4056 Basel

Professor Dr. Ekkehard Fluck, Institut für Anorganische Chemie der Universität Stuttgart, Pfaffenwaldring 55, D-7000 Stuttgart 80

Professor Dr. Robert Kosfeld, Institut für Physikalische Chemie der Rhein.-Westf. Technischen Hochschule Aachen, Tempelgraben 59, D-5100 Aachen

Editorial Board

Professor Sture Forsén, Department of Physical Chemistry, Chemical Centre, University of Lund, P.O.B. 740, S-22007 Lund

Professor Dr. Shizuo Fujiwara, Department of Chemistry, Faculty of Science, The University of Tokyo, Bunkyo-Ku, Tokyo, Japan

Dr. R. K. Harris, School of Chemical Sciences, The University of East Anglia, Norwich NR4 7TJ, Great Britain

Professor C. L. Khetrapal, Raman Research Institute, Bangalore – 560006, India

Professor E. Lippmaa, Department of Physics, Institute of Cybernetics Academy of Sciences of the Estonian SSR, Lenini puiestee 10, Tallinn 200 001, USSR

Professor G. J. Martin, Chimie Organique Physique, Université de Nantes, UER de Chimie, 38, Bd. Michelet, F-44 Nantes, B. P. 1044

Professor A. Pines, Department of Chemistry, University of California, Berkeley, CA 94720, USA

Professor Franz H. A. Rummens, Department of Chemistry, University of Regina, Regina, Saskatchewan S4S 0A2, Canada

Professor Dr. Bernard L. Shapiro, Department of Chemistry, Texas A and M University, College Station, TX 77843, USA

List of Contributors

Dr. Robin L. Armstrong, Department of Physics, University of Toronto, Toronto, Ontario M5S 1A7, Canada

Dr. Georges J. Bene, Département de Physique de la Matière Condensée, Section de Physique de l'Université de Genève, CH-Genève

Dr. R. Blinc, University of Ljubljana, J. Stefan Institute, Ljubljana, Yugoslavia and Solid State Physics Laboratory, ETH-Zürich, CH-Zürich

Dr. S. Clough, University of Nottingham, Great Britain

Dr. E. L. Hahn, Physics Department, University of California, Berkeley, CA 94720, USA

Dr. J. Jeener, Université Libre de Bruxelles, B-1050 Bruxelles

Dr. F. R. McCourt, Guelph-Waterloo Centre for Graduate Work in Chemistry and Departments of Chemistry and Applied Mathematics, University of Waterloo, Waterloo, Ontario, Canada

Dr. Milan M. Pintar, University of Waterloo, Faculty of Science, Department of Physics, Waterloo, Ontario N2L 3G1, Canada

Dr. Alfred G. Redfield, Departments of Physics and Biochemistry, and The Rosensteil Basic Medical Sciences Research Center, Brandeis University, Waltham, MA 02154, USA

Dr. J. S. Waugh, Department of Chemistry, Massachusetts Institute of Technology, Cambridge, MA 02139, USA

A Guide to Relaxation Theory

A. G. Redfield

Contents

Appendix A. Approach to Finite-Temperature Equilibrium
in a Semiclassical Theory 11
References 12

This lecture is designed to give a brief overview of relaxation theory. We will not repeat things which you can find in readily available texts and reviews, but merely point you to them. We give as classical a picture as possible, in the belief that this is most useful for experimentalists, and in the conviction that quantum versions of any successful classical picture can be made.

There are several recent reviews [1, 2] on this subject, as well as older, comparatively simple treatments [3, 4, 5, 6]. I have not recently worked in this area, so my treatment may contain some errors or omit important recent ideas.

Table 1. Some important interactions leading to relaxation, their magnitudes, and their correlation times. This list is not meant to be complete, and is included mainly to show the range of magnitudes and times

Measurement	Mechanism	Magnitude $\omega_i/2\pi$[Hz]	τ_c[sec]
$T_{1\rho}$ or T_2	Nuclear Spin-Spin Chemical Shift	1–10 10–1000	Nuclear T_1 Chemical Rate Consts.
T_1 or $T_{1\rho}$	Nuclear Dipolar Spin-rotational Electronic Dipolar Nuc Quadrupolar	$100-10^4$ $100-10^4$ 10^5-10^7 10^4-10^7	Rotational Dif'n Time τ_r Molec. Collision Time Faster of τ_r or electron T_1 τ_r, vibrations
EPR T_1	Hyperfine Anisotropy Crystal Field Anistropy	10^6-10^9 $0-10^{13}$	τ_r τ_r, vibrations

The Relaxation Mechanism. Table 1 lists a representative set of relaxation mechanisms, with their magnitudes expressed in frequency units as ω_i. Fig. 1 illustrates qualitatively what is meant by ω_i. The *mean square* interaction [ω_{pert}^2] is often, but not always, equal to [ω_i^2]. They are unequal when the interaction occurs only sparsely (Fig. 1b). The correlation time τ_c is also illustrated in Fig. 1. These definitions can only be made qualitatively [4].

The energy required in any relaxation process depends on what is being measured. I will denote it by ω_r. For T_1 and heteronuclear Overhauser measurements, ω_r is one or two times the resonance frequency, or the sums and differences between resonance frequencies of nearby dissimilar spins. For T_2 measurements, the contribution with $\omega_r = 0$ is dominant, as it is also for homonuclear Overhauser measurements. For $T_{1\rho}$ measurements, ω_r is generally one or two times γH_{eff}, where H_{eff} is the effective field in the rotating frame.

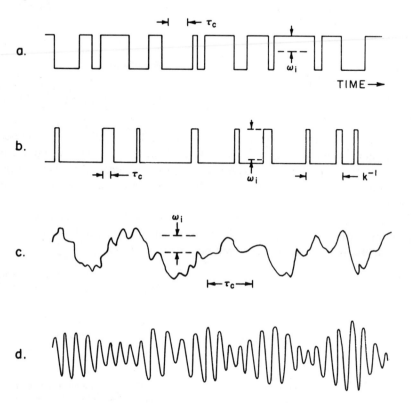

Fig. 1a–d. Some important kinds of random functions. The horizontal axis is time, and the vertical axis is an interaction energy which we express in frequency units. The quantity ω_i is roughly $[\langle\omega^4_{\text{pert}}\rangle/\langle\omega^2_{\text{pert}}\rangle]^{1/2}$. (a) A pertubation jumps between two values the probability per unit time of a transition being τ_c^{-1}. The correlation function of such a function is not exponential, but is a sum of Poisson distributions. Can you show this? (b) A "sparse" perturbation. The function jumps with a probability per unit time k to an unusual value, then jumps back with probability per unit time τ_c^{-1}. If $k \ll \tau_c^{-1}$, then such a perturbation has an exponential correlation function. (c) Many random functions look like this. If this is Johnson noise from a resistor and capacitor in parallel then it will have an exponential correlation time; it is also exponential for terms in the magnetic dipole-dipole interaction in a molecule undergoing isotropic rotational diffusion. Otherwise the correlation function is likely to be more complicated. (d) A sine wave undergoing random phase and amplitude variation. This type of noise is produced by an underdamped pendulum or LC circuit, or by a narrow band of lattice vibrations in a solid. Its correlation function is more or less a damped cosine function. Such a perturbation often cannot be treated in the ways described in the text, and may act via higher order (Raman) processes

The perturbation limit is the limit where the correct answer can be obtained by theory carried to second order in ω_{pert}. In this limit, as an approximation, we can write

$$T^{-1} = \sum \frac{[\omega_{\text{pert}}^2]\tau_c}{1 + \omega_r^2 \tau_c^2}, \qquad (1)$$

where T^{-1} represents the relaxation rate (T_1^{-1}, T_2^{-1}, $T_{1\rho}^{-1}$, or a cross-relaxation rate) and the summation sign indicates that a precise theory may require terms with a variety of frequencies ω_r, and terms from a variety of relaxation mechanisms with different ω_i and τ_c. Eq. (1) is approximate both with respects to magnitude and also with respect to its Lorentzian form. The latter is valid for important cases such as rotational diffusion, but not in general (e.g. dipolar relaxation via spatial diffusion).

When the relaxation mechanism is sparse, with collisions occuring at an average rate k sec^{-1} with a mean duration τ_c, then ω_{pert} [2] is zero most of the time and equals ω_i^2 for a fraction of the time $k\tau_c$. Its mean square strength is $\omega_i^2 k\tau_c$. Then (1) becomes

$$T^{-1} = \sum \frac{\langle \omega_i^2 \rangle \tau_c^2 k}{1 + \omega_r^2 \tau_c^2} \qquad (2)$$

Sometimes the time-varying perturbation can be divided into several terms of different time scales (i.e. vibration and tumbling) and these can be treated separately. Different terms may dominate for different measurements (i.e. T_1 and T_2) on the same system.

Relaxation times can be predicted by various semiclassical strategems. A random-walk model [7] in the rotating frame can be used to obtain T_2 [6]. T_1 can be expressed in terms of the Fourier transform of a correlation function using a very simple perturbation theory carried to second order [5]. $T_{1\rho}$, and its zero rf field limit T_2, can be calculated in exactly the same way by quantizing along the effective field. More formidable-looking density-matrix theories are simply generalizations of these ideas to more complicated systems. In general, such systems have no more spectral lines than would be predicted by considering all allowed and forbidden transitions, with relaxation rates porportional to spectral densities which are Fourier transforms of correlation functions of the perturbations.

A map of relaxation theory is shown in Fig. 2. The boundary between perturbation and strong collision limits occurs when the relaxation predicted by the perturbation limit for a single collision exceeds unity. This means that the predicted τ_c/T, for a dense perturbation (Fig. 1a), or the predicted $T^{-1}k^{-1}$, for a sparse perturbation, exceeds unity. In either case the boundary occurs when

$$\frac{\langle \omega_i^2 \rangle \tau_c^2}{1 + \omega_r^2 \tau_c^2} \approx 1 \qquad (3)$$

as indicated in Fig. 2.

Certain types of theories deal only with line-shapes and not with saturation [4, 8], or phenomena produced by strong pulses [9]. Such theories may be either simpler, or cover a different range of physical phenomena, than more general theories.

Sometimes one relaxation mechanism governs another: for example, collisions which govern relaxation of rotational energy of gas molecules, treated by strong collision theory, influence how spin-rotation interaction relaxes nuclei in the molecule. Or, the relaxation rate of an electron governs that of a nearly nucleus. Some caution should be used in these cases; it is safest to first consider the relaxation of the entire combined system (spin-rotation interaction plus nuclear spin; or electron plus nuclear spin) first, before concentrating on the nuclear spin.

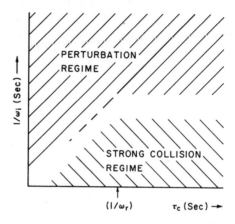

Fig. 2. A map of relaxation theory, showing where perturbation theory is applicable, and where strong collision theories must be applied.

Often expression (3) is much less than one for a T_1 process, where $\omega_r \approx 10^8$ for protons in a 10 kG field, but not for T_2, where $\omega_r = 0$ dominates. Then the perturbation limit can be used for T_1 but a static or slow motion method must be used for T_2 or $T_{1\rho}$.

The slow motion limit occurs when expression [3] is much less than one. The theory is complicated and cannot be generalized, i.e. a single theory cannot be given, even for a single system, that will cover different kinds of experiments (such as steady state saturation as opposed to pulsed). Generally the NMR spectrum is not simple even for a single spin, but reflects the distribution of ω_i viewed as a static perturbation. The dynamic behavior is complex. Often, when $\omega_r \cong 0$ then $T^{-1} \approx \tau_c^{-1}$.

An example of this is diffusional relaxation of nuclear dipole energy, where $T \equiv T_{1D} = \alpha \tau_c$, and α is only slightly less than one (see Ref. [9]). For Zeeman energy, which has a very high ω_r, on the other hand, T_1^{-1} will approach the perturbation limit.

Example. Consider the situation in Fig. 3a, where a spin spends most of its time resonating at ω_0 in a field H_0, but shifts occasionally, k times per second on the average, to a higher field where it resonates at $\omega_0 + \omega_i$, for an average time τ_c. An example would be a water proton which occasionally exchanges with an amide proton on a solute molecule. The mole fraction of solute to water proton sites (assuming no partially ionized groups) will be $k\tau_c$ and this is a perturbation of the type depicted in Fig. 1b. I will assume that $\omega_i \tau_c \gg 1$ so that the perturbation limit does *not* apply and there will be two lines observable: a strong one at ω_0, and one whose area is $k\tau_c$ smaller. These lines will be lifetime broadened, the weaker one with $T_2 \cong \tau_c$ and the stronger one narrower with a $T_2 = k^{-1}$. There is no fluctuating transverse field in this model, and therefore no T_1 process. We will assume such processes to occur very slowly by some other spin interaction which we will ignore except as a means to establish an initial magnetization.

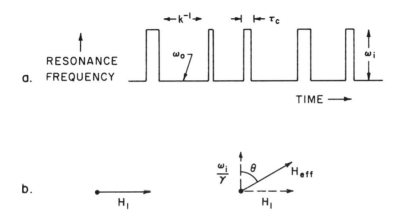

Fig. 3. (a) Time dependence of the resonance frequency of a water proton which occasionally exchanges with a solute proton. (b) Left, the effective field of a water proton if the rf frequency is at the water resonance. Right, the effective field experienced by the same spin in the same rf and magnetic field during the time of order τ_c, that it occupies a solute site

Following a classic treatment by Meiboom [10], let us predict the results of a $T_{1\rho}$ measurement of the water protons. The water proton spins are flipped 90° by a strong short pulse and then a weaker rf field is applied phase shifted by 90° so that the nuclear magnetization $\vec{M}$ is now aligned along H_1 in the rotating field. The rf frequency is equal to the water frequency so that the effective field that a spin experiences is most of the time transverse and equal to H_1, but occasionally flips up to the orientation shown on the right side of Fig 3b. Normally the water signal is monitored as a function of time and is observed to decay at a rate $T_{1\rho}$ proportional to the solute concentration and thus to k (I assume that the residence time on the solute, τ_c, is independent of solute concentration). Usually the solute concentration is so low that the signal due to protons on the solute can be ignored.

Consider some time t shortly after the start of the experiment, and focus on the small fraction of spins which started as water spins but are momentarily residing on solute. Call the magnetization this subset of spins had at the start $\vec{m}$. This magnetization was aligned along the H_1 direction. At the time t these spins have resided in a larger effective field (Fig. 3b, right side) for a short but *variable* time of order τ_c. If the angle θ defined in Fig 3b is not too small then the magnetization $\vec{m}'$ of this subset of spins at time t equals the projection of $\vec{m}$ on $\vec{H}_{eff}$, and $\vec{m}'$ is along $\vec{H}_{eff}$: $m' = m \cos \theta$. The orientation of each spin is invariant when the resonant frequency shifts suddenly, but the average transverse moment is zero because each spin has precessed a variable number of times, of the order of $\omega_i \tau_c / \sin \theta$ times around the new effective field.

This is a classical version of the treatment of antiadiabatic collisions in quantum mechanics. Note that it would be modified, but not simplified, if the flips of resonance frequencies were not sudden.

Now consider the situation at a time t', where $k^{-1} \gg t' - t \gg \tau_c$ (so that virtually all of this same subset of spin will have returned to a water site). The magnetization is again along H_1 and is reduced by another factor $\sin \theta$, so that $m \sin^2 \theta$. Thus the *change* in magnetization along H_1 is $(1 - \sin^2 \theta)$ times the magnetization, when one collision has occured on the average. If each collision completely randomized the spin orientation then the relaxation rate $T_{1\rho}^{-1}$ would equal the collision rate k. Instead there is only partial randomization, by a factor $1 - \sin^2 \theta = \cos^2 \theta$. Thus we expect

$$T_1^{-1} = k \cos^2 \theta = \frac{k \gamma^2 H_1^2}{\gamma^2 H_1^2 + \omega_i^2} . \tag{4}$$

This expression is compared with experiment to estimate ω_i and the residence time τ_c from a measurement of T_ρ as a function of H_1. Contributions of $T_{1\rho}$ due to other relaxation mechanisms must be estimated and subtracted. As indicated above, τ_c is k^{-1} times the mole fraction of solute (versus solvent) proton sites.

When γH_1 becomes sufficiently large the above strong sudden collison theory breaks down because, even though $\omega_i \tau_c \gg 1$, the change in the precession frequency about the effective field will be small:

$$[\omega_i - (\omega_i^2 + \gamma^2 H_1^2)^{1/2}] \tau_c \ll 1 , \tag{5}$$

When this occurs, however, perturbation theory is valid, and the theory is very closely analogous to calculating T_1 due to a small random perturbation except that H_1 replaces H_0 and the quantization axis is along H_1. Thus

$$T_{1\rho}^{-1} \cong \frac{\tau_c^2 \langle \omega_{pert}^2 \rangle}{1 + \gamma^2 H_1^2 \tau_c^2} = \frac{k \tau_c^2 \omega_i^2}{1 + \gamma^2 H_1^2 \tau_1^2} \tag{6}$$

To calculate (6) one uses the assumption that the correlation function is exponential:

$$\langle \omega_{pert}(t) \omega_{pert}(t + \tau) \rangle_t = k \tau_c \omega_i^2 e^{-|\tau|/\tau_c}.$$

The precise general solution to this problem, based on methods outlined below, is [*10*]

$$\frac{1}{T_{1\rho}} = \frac{k \tau_c^2 \omega_i^2}{1 + (\omega_i^2 + \gamma^2 H_1^2) \tau_c^2} \tag{7}$$

This is valid no matter the relative magnitudes of ω_i, γH_1, and τ_c^{-1}. It is consistent with both [4] and [6] in their respective ranges of validity.

The General Method of Coupled Equations of Motion. I illustrate this with a simple example, similar to the previous one. Suppose there are two otherwise equivalent conformations for a spin attached to a molecule, in which the spin feels two different magnetic fields $\vec{H}_A$ and $\vec{H}_B$ (Fig 4a). The conformation changes between A and B type with rates $k_{AB} = k_{BA} = k$, where k_{AB} is the kinetic rate constant for the transition from conformation A to conformation B.

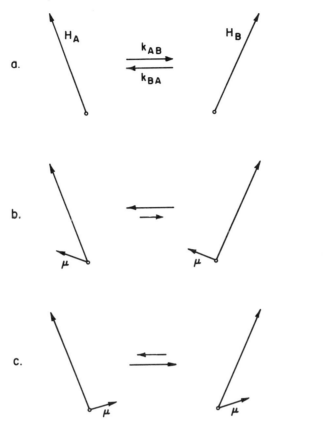

Fig. 4a–c. A molecule is assumed to shift between two conformations in which the field that it experiences differs. The field is assumed static in each conformation and is $\vec{H}_A$ in the A conformation and $\vec{H}_B$ in the B conformation, as shown in (a). A molecule whose spin has the orientation indicated by the vector $\vec{\mu}$ in (b) has lower magnetic energy in conformation A. The spin orientation does not change during the collision. A molecule whose spin has the orientation shown in (c) will prefer conformation B, where its energy is lower. The rates k_{ab} and k_{ba} in these cases are slightly different, as indicated by the horizontal arrows

This could be generalized as follows: The rates need not be equal, and many more than two states can be considered with interchange between any pair. The two fields $\vec{H}_A$ and $\vec{H}_B$ might be different simply because of a chemical shift difference, but more generally they represent a wide variety of more complicated interactions. For example they could represent the Hamiltonian for the electric quadrupole interaction of a spin greater than 1/2, which could change greatly in orientation if a molecule underwent a conformation change. Of course in that case the problem could not be described in terms of simple magnetic spin dynamics but the principles would be the same as long as appropriate correspondances are made. These are: $\vec{M}$ corresponding more or less to a density matrix; $\vec{H}_A$ and $\vec{H}_B$ to spin Hamiltonians; the spin magnetic moment to a wave function; Bloch's equation to a Liouville equation, and so on.

An appropriate way to treat this problem is to define magnetizations $\vec{M}_A$ and $\vec{M}_B$, which are associated with the pools of N_A and N_B spins residing on molecules in the A or the B conformations.

Appropriate equations, sometimes called the McConnel equations, are:

$$\frac{d\vec{M}_A}{dt} = \gamma(\vec{M}_A \times \vec{H}_A) + k\vec{M}_B - k\vec{M}_A \tag{8a}$$

$$\frac{d\vec{M}_B}{dt} = \gamma(\vec{M}_B \times \vec{H}_B) + k\vec{M}_A - k\vec{M}_B \tag{8b}$$

The first terms on the right represent the kinematic interaction of the spins with the fields $\vec{H}_A$ and $\vec{H}_B$, the second terms represent influx of magnetization from the other pool, and the third term represents outflow to the other pool. In general, $\vec{H}_A$ and $\vec{H}_B$ would have to include applied radio frequency fields, if any.

I will not attempt to discuss actual solutions to these equations; for an excellent early review see the article by C. S. Johnson [3], and for recent work see reference [2]. Most solutions of this class of equations deal with cases where the different coupled Hamiltonians commute with each other. (In terms of Fig 4, this means the fields $\vec{H}_A$ and $\vec{H}_B$ are collinear but of different magnitude.)

It is interesting to see what happens if this is not true, i.e. if $\vec{H}_A$ and $\vec{H}_B$ are not in the same direction. Then it is perhaps clear from our presentation of the Meiboom $T_{1\rho}$ theory that the spin magnetization would decay to zero after a long enough time, violating the reality of thermodynamic equilibrium. A standard way to avoid this problem is to arbitrarily replace the magnetization, which occurs in the relaxation terms in the equation of motion, by the deviation of the magnetization from a suitable equilibrium magnetization.

In the present case the procedure is to modify [8] to be

$$\frac{d\vec{M}_A}{dt} = \gamma(\vec{M}_A \times \vec{H}_A) + k(\vec{M}_B - \vec{M}_{B0}) - k(\vec{M}_A - \vec{M}_{A0}) \tag{9a}$$

$$\frac{d\vec{M}_B}{dt} = \gamma(\vec{M}_B \times \vec{H}_B) + k(\vec{M}_A - \vec{M}_{A0}) - k(\vec{M}_B - \vec{M}_{B0}) \tag{9b}$$

where $\vec{M}_{A0}$ and $\vec{M}_{B0}$ are the magnetizations that the pool of N_A and N_B spins will have in equilibrium in the fields $\vec{H}_A$ and $\vec{H}_B$ (for example, $\vec{M}_{A0} = N_A \vec{H}_A \mu^2 / kT$, where $\vec{\mu}$ is the nuclear moment, and the nuclear spin is 1/2). I believe that Eq. (9), and their straightforward analogues and generalizations provide a framework for solving all magnetic resonance relaxation problems, at least in the high temperature limit. In those cases where perturbation theories are not applicable, the equations must be solved by small or large scale computation. Only for the many body case (i.e. solids) is this likely to provide any fundamental difficulty.

In Appendix A, I give a justification of the alteration of (8) to (9), based on the following more fundamental appearing idea: Consider those molecules of the pools N_A and N_B which have their spin $\vec{\mu}$ oriented as in Fig 4b. The spin energy $-\vec{\mu} \cdot \vec{H}_A$ on the left is lower than that on the right, $-\vec{\mu} \cdot \vec{H}_B$, so this should be reflected in the kinetic constants as indicated in exaggerated form by the arrows for this class of spins. On the other hand, the reverse is true for molecules whose moment is as depicted in Fig 4c; for these the kinetic constant magnitudes will be reversed. When this microscopic idea is incorporated into a Boltzman equation for the spins, then the "hydrodynamic" Eqs. (9) follow (see Appendix A).

Memory Theories. Sometimes it is useful to view a spin as polarizing its surroundings, and to think of the extra interaction of this polarization as producing shifts in the NMR frequency and other effects. For example, in a metal, a nuclear spin can be said to polarize nearby electron spins, and this "clothed" moment (nuclear plus electron) combines to give a Knight-shifted resonance.

As a nuclear spin $\vec{\mu}$ precesses, these polarizations do not quite keep up with the spin, and this leads to a tiny field component $\vec{h}_f(t)$ not parallel to $\vec{\mu}$, so that $\vec{h}_f(t) \times \vec{\mu} \neq 0$ and the spin is affected by this field. This is a dissipative field and can presumably be related to the fluctuations that the spin feels by a local susceptibility and generalizations of the fluctuation-dissipation theorem; see the recent review by Kivelson and Ogan [1].

It is easy to see how such a field reaction also leads to relaxation to equilibrium in the perturbation limit: It is reasonable to assume that the reaction memory field points more or less where the spin was a short time ago, and deviates from the spin's direction by an amount proportional to the spin's angular velocity. Thus, the part $\vec{h}_f(t)$ which is perpendicular to $\vec{\mu}$ is $\vec{h}_f = -C \dfrac{d\vec{\mu}}{dt}$, where C is a constant which may depend on the surroundings. The extra term $\gamma(\vec{\mu} \times \vec{h}_f)$ in the microscopic equation of motion for the spins in thus $-C\gamma\left(\vec{\mu} \times \dfrac{d\vec{\mu}}{dt}\right) \approx -C\gamma[\vec{\mu} \times (\vec{\mu} \times \vec{H})]$. This vector expression has components perpendicular to $\vec{H}$ which would average to zero when an average over nuclear spin orientation is taken, but it always has a component along H which, when averaged, leads to an equilibrating term, i.e., the term $+M_0/T_1$ in the Bloch equations. Balancing this term is the term $-M/T_1$ which comes from the fluctuating part of the field. The friction constant C which was used above is determined by the requirement that the friction hydrodynamic term $-M_0/T_1$ must be of the right magnitude to give the correct equilibrium magnetization, just as the friction term in the Langevin equation for Brownian motion can be related [7] to the fluctuating force by the requirement that the mean kinetic energy is $3/2\, kT$.

Multistep Processes and Metalloprotein Geometry. Often a molecule or ion carrying a spin must undergo a two (or more) step process before that spin can relax. Consider the following reactions:

$$I\uparrow \rightleftharpoons I_1\uparrow \rightleftharpoons I_1\downarrow \rightleftharpoons I\downarrow, \tag{10}$$

$$E + I\uparrow \underset{k_2}{\overset{k_1}{\rightleftharpoons}} EI\uparrow \underset{k_4}{\overset{k_3}{\rightleftharpoons}} EI\downarrow \underset{k_1}{\overset{k_2}{\rightleftharpoons}} E + I\downarrow. \tag{11}$$

Eq. (10) can represent an Orbach process in which an ion I with spin up is thermally excited to a state I_1 with spin up. In this excited state the relaxation time may be rapid so the spin may flip over (third step) before I_1 returns to I. In Eq. (11) I may be a small molecule which binds to an enzyme E to form some kind of complex EI. If the enzyme E contains a magnetic ion it will rapidly relax a nuclear spin on the small molecule I as shown.

Eq. (11) is formally like the catalysis of an enzyme reaction; here the enzyme is catalyzing magnetic relaxation. The mathematics for analysis of such a process is virtually the same as that for elementary enzyme kinetics, yielding the Michaelis-Menten equation as derived in any elementary biochemistry text. The behavior is different depending on competition between the two rates k_2 and k_3, and on the concentration of I relative to the equilibrium constant k_2/k_1 for dissociation of EI.

The similarity between enzyme kinetics, and this type of nuclear spin relaxation (which is often applied to enzyme studies), does not seem to be widely realized. Measurements on such system are very useful in biochemistry because the rate

$$k_3 \cong k_4 \cong 2T_{1M}^{-1},$$

where T_{1M} is the T_1 that the spin would have if it were permanently bound to E, depends strongly on the distance of this spin from the magnetic center. Since T_{1M} is proportional to the distance to the sixth power, it does not require a very good experiment, and theory relating the experiment to distance, to get a good distance estimate. Accuracy of 10 % to 20 % is claimed by workers in this field, for distances in the 7 to 15 Å range. Knowledge of the distance of a spin on a bound molecule from the metal helps to elucidate enzyme mechanisms since such metals are often uniquely involved in catalysis.

This subject has been reviewed extensively [13], and I will simply list a number of pitfalls, most of which can be overcome with sufficient work and attention to detail:

1. If $k_2 \ll k_3 \approx k_4$ then the association or complexation step is rate determining (this is called "slow exchange") and the interesting magnetic rate $k_3 \approx k_4$ is not readily observable since, usually, the concentration of the complex EI is so low as to be unobservable. (I believe that this can be overcome by a careful study of saturation of the resonance of the species I by a radio frequency field whose frequency is far from I's resonance, acting indirectly via the broadened species EI.) Sometimes it is difficult to determine what is rate limiting; temperature dependence generally tells this since association and magnetic relaxation usually have opposite temperature dependences, but this is not completely reliable.

2. If the association is not rate limiting, then the experiment measures the time average of $\langle r^{-6} \rangle$, where r is the distance of the nuclear spin to the magnetic ion. There may be several association sites [symbolized by EI', EI'', etc. in Eq. (11)]. A nonproductive binding site closer to the ion than the normal one can then dominate the measurement even though it is occupied a small fraction of the time. On the other hand, the dominant long-lived site may be nonproductive, and the biochemically interesting site may be too shortlived to affect the measurement.

3. The electronic relaxation time τ_e must be known. Usually it is inferred from a series of nuclear relaxation measurements performed at various magnetic fields; the rate $k_3 \approx k_4$ decreases when $\gamma H_0 \tau_e \gtrsim 1$ and by suitable analysis τ_e can be inferred. Unfortunately τ_e itself varies with H_0, in many instances.

4. If the metal ion is not tightly bound to the enzyme it will also be free in solution, and this must be taken into account.

Thus, to do this type of experiment properly, relaxation measurements as a function of field, temperature, and concentration must be made. In addition the enzyme preparation must be properly characterized. The criteria for success are still not clear but an abviously necessary one is that distances measured to different nuclei on the same molecule permit building of an active site model without impossible bond lengths and angles. With the advent of high-field Fourier transform instruments such multiply determined geometries should be obtainable.

Appendix A. Approach to Finite-Temperature Equilibrium in a Semiclassical Theory

Apparently it is necessary to treat this problem with a Boltzmann equation. We define a probability density $P(\hat{\mu}_A)$ which is to be thought of as a function of the direction of a spin $\vec{\mu}_A$ in the A conformation; $\hat{\mu}_A$ is a unit vector in the $\vec{\mu}_A$ direction, and $P_A(\hat{\mu}_A)d\Omega_A$ is the number of spins whose direction is in the solid angle element $d\Omega_A$ about $\hat{\mu}_A$. P ist normalized: $\int P(\hat{\mu}_A)d\Omega_A = N_A$, where the integral is over all solid angles. A similar function $P_B(\hat{\mu}_B)$ is defined and normalized for spins in the B conformation.

A reasonable analog to (8a) and (8b) would be

$$\frac{dP_A(\hat{\mu}_A)}{dt} = \gamma(\vec{\mu}_A \times \vec{H}_A) \cdot \vec{\nabla}_A P_A + k_{BA} P_B(\hat{\mu}_A) - k_{AB} P_A(\hat{\mu}_A), \tag{A1a}$$

$$\frac{dP_B(\hat{\mu}_B)}{dt} = \gamma(\vec{\mu}_B \times \vec{H}_B) \cdot \vec{\nabla}_B P_B + k_{AB} P_A(\hat{\mu}_B) - k_{BA} P_B(\hat{\mu}_B). \tag{A1b}$$

The last two terms of each equation represent the flux of spins occurring as A molecules become B molecules as a result of conformation changes. The first terms are kinematic terms similar to the kinematic terms

$$\vec{v} \cdot \nabla_r \rho(\vec{p}, \vec{r}) + \vec{f} \cdot \nabla_p \rho(\vec{p}, \vec{r})$$

which occur in the Boltzmann equation for gases, where $\vec{v}$ is the velocity, $\vec{f}$ the force, ρ the probability density in real and momentum space, and $\vec{\nabla}_r$ and $\vec{\nabla}_p$ are its partial gradients. In (A1), the symbols ∇_A and ∇_B represent angular gradients of P_A and P_B.

I now introduce the only new idea. As I suggested in the main text in connection with Fig. 4, the rate of transfer of a set of molecules between A and B will be microscopically influenced by the orientation of their spins. Processes which gain (spin) energy will go more slowly than those which lose it:

$$k_{AB}/k_{BA} = e^{-\vec{\mu} \cdot (\vec{H}_B - \vec{H}_A)/kT}. \tag{A2}$$

Thus in (A1a), it is reasonable that

$$k_{AB} = k e^{-\vec{\mu}_A \cdot (\vec{H}_B - \vec{H}_A)/2kT} \tag{A3a}$$

$$k_{BA} = k e^{-\vec{\mu}_A \cdot (\vec{H}_A - \vec{H}_B)/2kT} \tag{A3b}$$

where k is an average rate constant ignoring spin effects. Eqs. (A3) are consistent with (A2).

In Eq. (A1b) a similar substitution would be made except that, of course, $\vec{\mu}_B$ would replace $\vec{\mu}_A$:

$$\frac{dP_A(\hat{\mu}_A)}{dt} = \gamma(\vec{\mu}_A \times \vec{H}_A) \cdot \vec{\nabla}_A P_A + k[e^{-\vec{\mu}_A \cdot (\vec{H}_A - \vec{H}_B)/2kT} p_B(\hat{\mu}_A) - \\ - e^{-\vec{\mu}_A \cdot (\vec{H}_B - \vec{H}_A)/2kT} p_A(\hat{\mu}_A)] \tag{A4a}$$

$$\frac{dP_B(\hat{\mu}_B)}{dt} = \gamma(\vec{\mu}_B \times \vec{H}_B) \cdot \vec{\nabla}_B P_B + k[e^{-\vec{\mu}_B \cdot (\vec{H}_B - \vec{H}_A)/2kT} p_A(\hat{\mu}_B) + \\ + e^{-\vec{\mu}_B \cdot (\vec{H}_A - \vec{H}_B)/2kT} p_B(\hat{\mu}_B)] . \tag{A4b}$$

Having set up a reasonable Boltzmann equation, we turn to its solution. Within the high-temperature approximation a reasonable trial solution is

$$P_A = \frac{1}{4\pi} N_A + \frac{3}{4\pi} \frac{\vec{M}_A \cdot \vec{\mu}_A}{\mu^2} . \tag{A5}$$

Here N_A, μ, and $\vec{M}_A$ have the same meaning they had before. You can verify that $\vec{M}_A$ is in fact the magnetization of the A spins from the definition of the components of M_A, for example the Z component: $M_{AZ} = \int \mu_A \cdot z\, P_A(\mu_A) d\Omega_A$. To show this rewrite this integral and (A5) in polar coordinates with axis along the z direction, and evaluate.

To get the "hydrodynamic" Eq. (9), substitute (A5) and its analog for P_B into (A4), and in the latter replace all exponentials by their high-temperature expansions. The rest is straight-forward algebra, which is easier than it may appear.

References

[1] Kivelson, D., Ogan, K.: Advan. Magnet. Res. 7, 72 (1974).
[2] Muus, L., Atkins, P.: Electron Spin Relaxation in Liquids. New York: Plenum Press 1972.
[3] Johnson, C. S. jr.: Advan. Magnet. Res. 1, 33 (1965).
[4] Slichter, C. P.: Principles of Magnetic Resonance. New York–London: Harper and Row 1963.
[5] Poole, C. P., Farach, H.: Relaxation in Magnetic Resonance. New York: Academic Press, (1971).
[6] Schumacher, R. T.: Introduction to Magnetic Resonance. New York: W. A. Benjamin 1970.
[7] Reif, F.: Statistical and Thermal Physics. New York: McGraw Hill, (1965).
[8] Van Vleck, J. H.: Phys. Rev. 74, 1168 (1948).
[9] Ailion, D.: Advan. Magnet. Res. 5, 177 (1971).
[10] Meiboom, S.: J. Chem. Phys. 34, 375 (1961).
[11] Redfield, A. G.: Advan. Magnet. Res. 1, 1 (1965).
[12] Dwek, R. A.: NMR in Biochemistry. Oxford: 1973.

Thermodynamics of Spin Systems in Solids. An Elementary Introduction

J. Jeener

Contents

Grand Ensemble Method 15
A Simple Example 17
The Approximation of Weak Order or High Temperature 18
Line Shapes (Weak Order) 19

Thermodynamics (or, more exactly, Thermostatics) and equilibrium statistical mechanics deal with N-body systems at equilibrium. When the nature of the N-body system is known (known Hamiltonian and commutation properties of the relevant dynamical variables for instance), each of its equilibrium situations can be specified in a very simple way by the values of a small number of macroscopic quantities: the analytical invariants of the motion of the system. As soon as the entropy is known as a function of these invariants, all other thermodynamic properties can be evaluated directly.

Before using thermodynamics or equilibrium statistical mechanics, one has to solve two preliminary and related problems:
– which are the relevant invariants of the motion?
– how long do we have to wait for equilibrium?
In most situations of interest, these two problems cannot be solved from first principles, so that one has to supplement theoretical results with empirical information derived from experiment.

In order to illustrate the above remarks, let us examine a few typical examples:

Dilute He^4 gas. The relevant invariants are (1) total energy, (2) number of particles, (3) total momentum and total angular momentum. In the situations typical of chemical physics, which we are interested in here, the invariance of momentum and angular momentum is suppressed by the box containing the gas. The evolution towards equilibrium is well understood from the kinetic theory of gases: homogeneous deviations from equilibrium decay in a time comparable to the mean time between two collisions on one atom, whereas inhomogeneous deviations (such as a temperature gradient) decay in much longer times, usually proportional to the square of the spatial extension of the inhomogeneity. In this conventional discussion of He^4 gas, we have taken a number of experimental observations for granted: helium does not form highly stable molecules (the number of which would be an independent invariant), nor do its nuclei break into lighter nuclei or assemble in heavier nuclei in usual conditions.

Dilute Hydrogen Gas. In the usual conditions of chemical physics, obvious invariants are (1) total energy and (2) the number of H_2 molecules. If one tries to describe experiments which do not last many days and are performed in the absence of suitable catalysts, ortho-para conversion is such a slow process that the reasonable attitude is to consider the number of ortho-H_2 molecules (or para-H_2 molecules) as an additional independent invariant (the total number of H_2-molecules is the sum of the numbers of ortho and para molecules, so that out of these three numbers, only two are *independent* invariants). In the case of experiments performed in still shorter times, one might wonder whether the numbers of molecules in each rotation-vibration level should not be considered as independent invariants in some extreme situations. Of course, if hydrogen is subjected to an external magnetic field, the energy of interaction of the proton spins with this field ("Zeeman energy") will also behave as an independent invariant for times shorter than the usual spin-lattice relaxation time T_1.

Spins 1/2 in a Rigid Lattice in Zero External Field. The total spin energy, which originates from spin-spin couplings, seems to be the only relevant invariant in this case. Spatially homogeneous deviations from equilibrium decay in times of the order of T_2, whereas inhomogeneous distributions of spin-spin energy relax much more slowly by the transport process known as spin diffusion.

One Species of Spins 1/2 in a Rigid Lattice, in a Large External Magnetic Field. This is the system on which we shall focus attention in the present lectures. Of course, total spin energy is an exact invariant of the motion because the Hamiltonian is independent of time. This total spin energy E can be expressed as a sum of two terms: the energy E_0 of coupling of the spins with the external field H_0 and the energy E_{ss} of coupling of the spins between themselves. Whenever the external field H_0 is much larger than the local field H_L (roughly speaking, H_L is the r.m.s. field caused by its neighbours at the location of any given spin), E_0 and E_{ss} also behave as invariants of the motion. This experimental result can be understood in the following simple way: the smallest amount by which a quantum process can change E_0 is $\pm \gamma \hbar H_0$, corresponding to the flip of a single spin in the external field (γ is the magnetogyric ratio of the spin). This process must conserve total spin energy, so that the change in E_0 must be compensated for by a rearrangement of spins causing an exactly opposite change in E_{ss}. Interchanging the states of two spins changes E_{ss} by an amount of the order of $\gamma \hbar H_L$, so that the number of spins involved in the necessary rearrangement must be, at least, of the order of H_0/H_L. If this number is large (ten or more), the processes which could change E_0 and E_{ss} are so complicated that their probability is negligibly small.

The invariants E, E_0 and E_{ss} are related by $E = E_0 + E_{ss}$, so that only two of them are *independent* invariants of the motion.

Disregarding total momentum, we note that, in all cases, one invariant (total energy or E_{ss}) can vary in a continuous way, whereas all the other invariants can only change by integral multiples of some basic jump (E_0, numbers of molecules, occupation number of some single molecule quantum state, . . .).

Grand Ensemble Method [*]

Let us now briefly review the grand ensemble method for evaluating the density operator of a system in equilibrium in the presence of a number of independent invariants of the motion, and the relation between entropy and the invariants.

We shall denote the invariants by $A, B, \ldots, E$, where E is the total energy, and assume for simplicity that all those invariants commute with each other. If we use the microcanonical ensemble point of view, we shall assume that our system has equal probabilities of being in any of the $\Gamma(A, B, \ldots, E)$ simultaneous eigenstates of the various invariants having eigenvalues in the ranges $(A, A + \Delta A), (B, B + \Delta B), \ldots, (E, E + \Delta E)$, and a probability of zero of being in any other state. The standard discussion of the microcanonical ensemble shows that the quantity $S(A, B, \ldots, E)$ given by

$$S(A, B, \ldots, E) = k \cdot \log \Gamma(A, B, \ldots, E),$$

where k is Boltzmann's constant, has all the properties of entropy as discussed in conventional equilibrium thermodynamics. However, in spite of its apparent simplicity, we shall not use this method for the evaluation of thermodynamic and spectroscopic properties, because the discontinuous variation of the occupation probability of a state as a function of the invariants make the calculations almost intractable.

A convenient solution to these problems is provided by the "grand ensemble" point of view, in which one evaluates the equilibrium properties of a system (denoted here by a subscript 1) which is supposed to be weakly coupled to a much larger system (denoted here by a subscript 2) in such a way that the two systems can slowly exchange the various relevant invariants. We shall assume that the combination of the two systems is in equilibrium, and that this combination has invariants in the ranges $(A, A + \Delta A)$, $(B, B + \Delta B), \ldots, (E, E + \Delta E)$. Any eigenstate of the various invariants for the combination $(1 + 2)$ then has the same occupation probability $1/\Gamma(A, B, \ldots, E)$. In the limit of very weak coupling between systems 1 and 2, these eigenstates for $(1 + 2)$ can be chosen as products of an eigenstate of system 1 and an eigenstate of system 2, with each invariant for $(1 + 2)$ equal to the sum of the corresponding invariants for 1 and for 2.

Let us now focus our attention on a single eigenstate of system 1, with the eigenvalues $A_1, B_1, \ldots, E_1$ of the invariants. If system 1 is indeed in this state, the invariants for system 2 will be in the ranges $(A - A_1, A - A_1 + \Delta A), (B - B_1, B - B_1 + \Delta B)$, $\ldots, (E - E_1, E - E_1 + \Delta E)$, and the number of allowed states for system 2 will be $\Gamma_2(A - A_1, B - B_1, \ldots, E - E_1)$. We can now evaluate the probability P_1 of finding system 1 in the above completely specified quantum state as the product of the number of corresponding states of system 2 and the probability of any allowed state of the combination system:

$$P_1(A_1, B_1, \ldots, E_1) = \Gamma_2(A - A_1, B - B_1, \ldots, E - E_1)/\Gamma(A, B, \ldots, E).$$

---

[*] See for instance K. Huang, "Statistical Mechanics", Wiley, New York, 1963, or R. Balescu, "Equilibrium and non-equilibrium statistical mechanics", Wiley, New York, 1975

It is tempting, at this point, to argue that the quantities $A_1, B_1, \ldots, E_1$ for the small system 1 are always very small compared to quantities $A, B, \ldots, E$ for the much larger system 2, in order to replace $\Gamma_2(A - A_1, \ldots)$ by the first few terms of a series expansion in powers of $A_1, B_1, \ldots, E_1$. Closer examination of this proposal shows that Γ and its derivatives are such rapidly varying functions of their arguments that a very large number of terms of the series expansion would be necessary for the usual range of equilibrium fluctuations of the invariants of system 1; however this difficulty can be resolved by taking the log before expanding as a power series with the following result:

$$\log P_1(A_1, B_1, \ldots, E_1)$$
$$= K - A_1 \, \partial \log \Gamma_2(A, B, \ldots, E)/\partial A - \ldots$$
$$- E_1 \, \partial \log \Gamma_2(A, B, \ldots, E)/\partial E$$
$$+ \text{higher oder terms}$$

where K does not depend upon $A_1, B_1, \ldots, E_1$. Using the known relations between S and Γ and neglecting higher order terms we can write the above expression under the more familiar form

$$P_1(A_1, B_1, \ldots, E_1) = \frac{1}{\Xi} \, e^{-A_1 \frac{1}{k} \left(\frac{\partial S}{\partial A}\right)_2 - \ldots - E_1 \frac{1}{k} \left(\frac{\partial S}{\partial E}\right)_2},$$

where the partial derivatives $(\partial S/\partial A)_2, \ldots$, are properties of the large system 2 (usually called thermostat) alone, and the grand partition function Ξ is a normalization constant which only depends upon these partial derivatives. We immediately notice that $(\partial S/\partial E)_2$ is nothing else than the inverse of the temperature of the thermostat, so that the E_1 dependence of P_1 is given by the usual Boltzmann factor $\exp(-E_1/kT)$, as expected. Similarly, if A is the number of molecules of some kind in the system, then the coefficient of A_1 in the above exponential is proportional to the usual chemical potential of these molecules.

If system 1 itself is a macroscopic system in equilibrium, the fact that it is weakly coupled with the larger system 2 will only cause the invariants $A_1, B_1, \ldots, E_1$ to undergo relatively small fluctuations around their average values, so that system 1 will also have a well defined entropy, temperature, chemical potentials, ... One can further show that temperature and chemical potentials will be the same for systems 1 and 2 at equilibrium, and that the relations between entropy and the invariants as predicted by the microcanonical ensemble or by the grand ensemble methods are the same.

Let us now introduce the quantum operators $\mathcal{A}, \mathcal{B}, \ldots, \mathcal{E}$ corresponding to the invariants $A, B, \ldots, E$ for system 1. We have already assumed that these operators commute with one another. The density operator describing the equilibrium situation of the macroscopic system 1 weakly coupled to the large thermostat 2 is a diagonal operator in a representation which diagonalizes the invariants $\mathcal{A}, \mathcal{B}, \ldots$ and $\mathcal{E}$, with diagonal matrix elements giving the occupation probability of the relevant state:

$$\rho_{eq} = \frac{1}{\Xi} \, e^{-\mathcal{A} \frac{1}{k} \left(\frac{\partial S}{\partial A}\right) - \ldots - \mathcal{E} \frac{1}{k} \left(\frac{\partial S}{\partial E}\right)}.$$

As we have noted, the partial derivatives such as $(\partial S/\partial A)$ appearing in the above expression for ρ_{eq} can be considered as properties either of the large thermostat 2 or of the system 1 in which one is directly interested. We shall usually take this second point of view and not mention the thermostat 2 anymore. We can then (at least in principle) use the relation $\langle A \rangle = \text{tr}(A \cdot \rho)$ to evaluate the average value of any invariant as a function of the partial derivatives of S. When this done for all invariants, one can invert this set of functions in order to express the partial derivatives of S as functions of the average values of the invariants, and integrate in order to express the entropy S itself as a function of the average values of the invariants. In the case of spin systems, in the usual limit of weak order, the evaluation of entropy in this way only involves elementary (even trivial) calculations.

A Simple Example

We shall now discuss some properties of an assembly of a large number of spins, all of the same species, located on a simple rigid lattice and subjected to a constant magnetic field H_0 much larger than the local field H_L. We shall make the further simplifying assumption that no quadrupolar effects are present. The Hamiltonian $\mathcal{H}$ can be written as

$$\mathcal{H} = \mathcal{H}_0 + \mathcal{H}_{ss} \text{ where } \begin{array}{l} \mathcal{H}_0 = \hbar\omega_0 \sum_i I_{Zi} = \hbar\omega_0 I_Z \\ \mathcal{H}_{ss} = \mathcal{H}' + \mathcal{H}'' \end{array}$$

where ω_0 is the NMR frequency under the influence of H_0 alone, I_{Zi} is the Z-component of the spin operator $\vec{I}_i$ for spin i, the summation goes over all spins, and the spin-spin coupling Hamiltonian $\mathcal{H}_{ss}$ has been split into its "secular" part $\mathcal{H}'$ which has nonzero matrix elements only between states with the same eigenvalue of $\mathcal{H}_0$, and its "non secular" part $\mathcal{H}''$ which connects states with different eigenvalues of $\mathcal{H}_0$.

Let us first discuss the energy levels of this system. If the external field H_0 is much larger than the local field H_L, we anticipate that a perturbative approach (with $\mathcal{H}_{ss}$ as the perturbation) will be suitable. The phenomena which we shall discuss in the present lectures can be quantitatively predicted by using first order perturbation only, so that the non-secular term $\mathcal{H}''$ plays no role here and the Hamiltonian can be approximated by

$$\mathcal{H}_{\text{effective}} = \mathcal{H}_0 + \mathcal{H}'$$

Let us denote by $|M, n\rangle$ the simultaneous eigenstates of the commuting operators I_Z and $\mathcal{H}'$:

$I_Z |M, n\rangle = M | M, n\rangle$

$\mathcal{H}' |M, n\rangle = E'(M, n) | M, n\rangle$

---

*) See M. Goldman, "Spin temperature and nuclear magnetic resonance in solids", Clarendon Press, Oxford, 1970 and references mentioned there. See also J. Jeener, Adv., in Magnetic Resonance. 3, 205, 1968).

where M is an integer (or a half-integer in the case of an odd number of half integer spins) and n denotes all the other quantum numbers which are required to completely specify each state.

If we exactly followed the scheme as outlined above ("grand ensemble method"), our two commuting and independent invariants would be total energy $\langle\mathcal{H}\rangle$ and the half difference between the numbers of up and down spins $\langle I_Z\rangle$. Further discussions are slightly simplified by a different choice of independent invariants: $\langle I_Z\rangle$ and $\langle\mathcal{H}'\rangle = \langle\mathcal{H}\rangle - \hbar\omega_0\langle I_Z\rangle$. The equilibrium density operator can then be written as

$$\rho_{eq} = \frac{1}{\Xi} e^{-\alpha I_Z - \beta \mathcal{H}'}$$

where $\Xi = \mathrm{tr}(e^{-\alpha I_Z - \beta \mathcal{H}'})$ and $\beta = \frac{1}{k}\frac{\partial S}{\partial \langle \mathcal{H}'\rangle}$

plays the usual role of $1/kT$ in more conventional systems, and α plays the role of a chemical potential.

The Approximation of Weak Order or High Temperature

The situation of complete spin disorder, in which all quantum states of the spin system are equally probable, is described by the value zero of the parameters α and β in the above expression of ρ. In most situations of practical interest in NMR and even in EPR, the maximum range of the energy of coupling of a single spin with external or internal fields is very much smaller than kT_L (where T_L is the lattice temperature), so that one anticipates that the equilibrium situations of the spin system will never deviate much from complete disorder and that its properties will be well approximated by the first non-trivial term of their expansion as a power series in α and β.

For the evaluation of properties such as the invariants and the rf susceptibility, which are linear in α and β for small values of these parameters, we can then replace the "exact" expression of the equilibrium density operator by an expansion as a power series in α and β, limited to linear terms in α and β:

$$\rho_{eq} = \frac{1 - \alpha I_Z - \beta\mathcal{H}' + \frac{1}{2}(-\alpha I_Z - \beta\mathcal{H}')^2 + \ldots}{\mathrm{tr}\{1 - \alpha I_Z - \beta\mathcal{H}' + \frac{1}{2}(-\alpha I_Z - \beta\mathcal{H}')^2 + \ldots\}}$$

$$\cong \frac{1}{\mathrm{tr}(1)}(1 - \alpha I_Z - \beta\mathcal{H}').$$

In this approximation of weak order, the average values of the invariants are given by (the operators I_Z and $\mathcal{H}'$ are traceless and orthogonal to each other):

$$\langle\mathcal{H}'\rangle = \mathrm{tr}\{\mathcal{H}' \cdot \rho_{eq}\} = -\beta(\mathrm{tr}\,\mathcal{H}'^2/\mathrm{tr}\,1),$$

$$\langle I_Z\rangle = \mathrm{tr}\{I_Z \cdot \rho_{eq}\} = -\alpha(\mathrm{tr}\,I_Z^2/\mathrm{tr}\,1),$$

the relations $\alpha = (1/k)\,(\partial S/\partial \langle I_Z\rangle)$ and $\beta = (1/k)\,(\partial S/\partial \langle \mathcal{H}'\rangle)$ which define α and β can be integrated easily from the situation of complete disorder $\alpha = 0, \beta = 0$

$$S(\alpha, \beta) - S(\text{disorder})$$

$$= -\alpha^2 \cdot \frac{1}{2} k\,(\mathrm{tr}\,I_Z^2/\mathrm{tr}\,1) - \beta^2 \cdot \frac{1}{2} k\,(\mathrm{tr}\,\mathcal{H}'^2/\mathrm{tr}\,1)$$

$$= -\langle I_Z\rangle^2 \cdot \frac{1}{2} k\,(\mathrm{tr}\,1/\mathrm{tr}\,I_Z^2) - \langle \mathcal{H}'\rangle^2 \cdot \frac{1}{2} k\,(\mathrm{tr}\,1/\mathrm{tr}\,\mathcal{H}'^2)$$

and the total spin energy E can be written as

$$\langle E\rangle = \hbar\omega_0 \langle I_Z\rangle + \langle \mathcal{H}'\rangle.$$

We note that, in the present case of a very simple spin system, in the limit of weak order, the thermodynamics is that of a juxtaposition of two completely independent subsystems: the spin-spin subsystem with energy $\langle \mathcal{H}'\rangle$ and a contribution to total entropy proportional to $\langle \mathcal{H}'\rangle^2$, and the Zeeman subsystem with energy $\hbar\omega_0\langle I_Z\rangle$ and an entropy contribution in $\langle I_Z\rangle^2$. This remark has made it very natural to also associate a temperature to each of those subsystems: the "dipolar temperature" $T_D = 1/\beta k$ is the temperature which we have already introduced, and the "Zeeman temperature" $T_Z = \hbar\omega_0/\alpha k$ is, as usual, the derivative of Zeeman energy with respect to Zeeman entropy. However, outside the limit of weak order the two subsystems are not independent of each other any more and the definition of two temperatures in the usual sense becomes ambiguous. Similar difficulties arise, even in the case of weak order, for spins with unequally spaced energy levels. Moreover, when spin subsystems exchange energy, the process often conserves other invariants, and this prevents the temperatures from equalizing in equilibrium as anticipated from our usual experience with temperatures. As a conclusion, I feel that it is more appropriate to introduce only one "temperature" and a number of parameters of the type α, analogous to chemical potentials.

Line Shapes (Weak Order)

The observed line shape for the absorption of rf energy at some frequency ω can be evaluated as the product of two frequency dependent quantities: one is the average probability of all transitions involving an increase of the total spin energy by $\hbar\omega$, and the other one is the difference in occupation probability between those states.

The intense NMR absorption line centered around the Larmor frequency ω_0 of the spins corresponds to individual processes in which the quantum number M changes by ± 1 (this changes the spin energy by $\pm \hbar\omega_0$) and the spin-spin coupling energy changes by an amount comparable to the coupling of one spin with its neighbours (this results in a line width of the order of γH_L).

Let us first assume that our spin system has "purely Zeeman order", which means $\alpha \neq 0$ and $\beta = 0$. The relevant population difference between a lower energy state $|M, a\rangle$ and a higher energy state $|M + 1, b\rangle$ is $\alpha/\mathrm{tr}\,1$ and depends only on the difference in M between the states as shown in the left hand part of Fig. 1. This population difference does not depend upon the spin-spin coupling energies so that we predict that the line will be in absorption over its whole width (in the case $\alpha > 0$), and we can write:

$$\chi_Z''(\omega) = \alpha \cdot f(\omega - \omega_0)$$

where $\chi_Z''(\omega)$ is the absorption part of the rf susceptibility for $\beta = 0$, and f describes the frequency variation of the average transition probability.

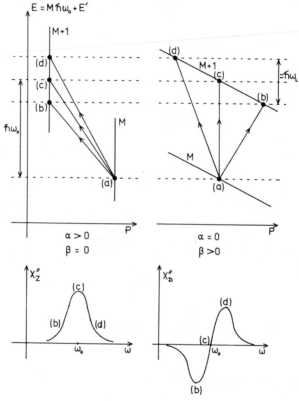

Fig. 1. Upper part: relation between total spin energy E and occupation probability P of individual quantum states. Lower part: frequency dependence of the absorption susceptibility χ''. Left hand side: Zeeman order $\alpha > 0$, $\beta = 0$ ($\omega_0 > 0$). Right hand side: Dipolar order $\alpha = 0$, $\beta > 0$

Let us now examine the complementary case of "purely spin-spin order" in which $\alpha = 0$ and $\beta \neq 0$. The relevant population difference between states $|M, a\rangle$ and $|M + 1, b\rangle$ is now $\beta(E_b' - E_a')/\mathrm{tr}\, 1$ and depends only on the difference in spin-spin energy between the states as shown in the right hand part of Fig. 1. We see that there will be no absorption at the "exact" NMR frequency ω_0 because this corresponds to the same spin-spin energy for the two states and, thus, to no difference in population. We also see that the line will be in absorption on one side of ω_0 and in emission on the other side, and that we can write:

$$\chi_D''(\omega) = \beta \cdot \hbar(\omega - \omega_0) \cdot f(\omega - \omega_0).$$

If both α and β are different from zero, the susceptibility is the sum of the contributions from α and β:

$$\chi''(\omega) = \chi''_Z(\omega) + \chi''_D(\omega)$$

When a spin system is in complete equilibrium with the lattice at temperature T_L, $\beta = 1/k\,T_L$ and $\alpha = \hbar\omega_0/kT_L$. Using the above expressions for χ''_Z and χ''_D, and noting that the line shape function $f(\omega - \omega_0)$ decreases very rapidly whenever $|\omega - \omega_0|$ is appreciably larger than $\omega_L = \gamma H_L$, we conclude that for such a spin system in complete equilibrium, the observed susceptibility is dominated by $\chi''_Z(\omega)$, whereas $\chi''_D(\omega)$, which is of the order of H_0/H_L smaller, is too weak to be observed in practice.

In order to observe the thermodynamic and spectroscopic effects of spin-spin order (i.e. of the fact $\beta \neq 0$) one has to artificially increase $|\beta|$, and this can be done in practice by various tricks which increase $|\beta|$ while decreasing $|\alpha|$. An upper limit to the possible increase of $|\beta|$ is set by the fact that, at best, spin entropy is conserved by the "trick". Adiabatic demagnetization in the rotating frame, irradiating the wing of the line, two phase shifted pulses ... are such tricks which either almost conserve entropy or are not much poorer at generating large values of $|\beta|$.

In the case of dipolar coupling between spins, if the Zeeman and dipolar contributions to spin entropy are equal, the maximum values of χ''_Z and χ''_D are comparable. As a conclusion, if an ordinary (Zeeman) NMR signal can be seen, then, using the proper tricks, the corresponding dipolar signal can also be seen with a comparable signal-to-noise ratio.

Coherent Averaging and Double Resonance in Solids

J. S. Waugh

Contents

1. Coherent Averaging 23
2. Frames of Reference: Truncation 25
3. Dipolar Narrowing 27
4. Other Internal Hamiltonians 28
5. Decoupling of Unlike Spins 28
6. Recoupling of Unlike Spins 29
7. Dipolar Oscillations 30
8. Alchemy 30

1. Coherent Averaging

In transient NMR experiments we are concerned with the expectation values of operators Q at some time t, $\langle t|Q|t\rangle$, in a system which has developed from some initial state $|0\rangle$. Thus we must solve the problem posed by

$$|t\rangle = U(t, 0)|0\rangle \tag{1}$$

where the propagator U is governed by the time-dependent Schrödinger equation

$$\overset{\circ}{\psi} = \lambda \mathcal{H} \psi \, ; \, \lambda = -\,i/\hbar. \tag{2}$$

If $\mathcal{H}$ is independent of time,

$$U = \exp(\lambda \mathcal{H} t). \tag{3}$$

If not, break up t into small intervals τ during which $\mathcal{H}$ is sensibly constant

$$U(N\tau) = \prod_{n=0}^{N-1} \exp[\lambda \mathcal{H}(n\tau)\tau] \tag{4}$$

In the limit $\tau \to 0, N \to \infty, N\tau = t$, this may be written

$$U(t) = T \exp[\lambda \int_0^t dt' \, \mathcal{H}(t')] \tag{5}$$

The time-ordering symbolized by T is essential if $[\mathcal{H}(t'), \mathcal{H}(t'')] \neq 0$ for any t', t'' in $\{0, t\}$

It is convenient to express the time-dependent problem (5) in terms of an equivalent time-independent one, replacing $\mathcal{H}(t)$ by some constant effective Hamiltonian $\overline{\mathcal{H}}$ which would have brought the system to $|t\rangle$ from $|0\rangle$ by some different path. $\overline{\mathcal{H}}$ is in some sense an "average" of $\mathcal{H}(t)$ over that interval

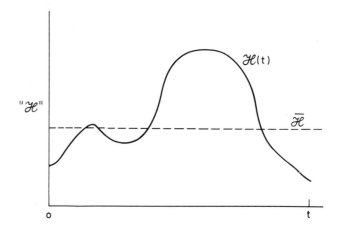

A recipe for constructing $\bar{\mathcal{H}}$ given $\mathcal{H}(t)$, is given by the Magnus formula, which yields $\bar{\mathcal{H}}$ as a power series in λ. The terms in the series can be evaluated by comparing a moment expansion of U

$$U = 1 + \lambda \int_0^t \mathcal{H}(t')dt' + \lambda^2 \int_0^t \int_0^{t'} dt' \, dt'' \, \mathcal{H}(t') \, \mathcal{H}(t'') + \ldots \tag{6}$$

with a cumulant expansion:

$$U = \exp(\lambda \bar{\mathcal{H}} t)$$

$$\bar{\mathcal{H}} = \sum_{n=0}^{\infty} \bar{\mathcal{H}}^{(n)} = \frac{1}{t}(a + \lambda b + \lambda^2 c + \ldots). \tag{7}$$

Exercise: Verify that

$$\bar{\mathcal{H}}^{(0)} = \frac{1}{t} \int_0^t \mathcal{H}(t')dt', \tag{8}$$

$$\bar{\mathcal{H}}^{(1)} = \frac{-i}{2t} \int_0^t dt' \int_0^{t'} dt'' [\mathcal{H}(t'), \mathcal{H}(t'')]. \tag{9}$$

Of course $\bar{\mathcal{H}}$ depends parametrically on the upper limit t. This awkward feature is changed if $\mathcal{H}(t)$ is periodic modulo some "cycle time" t_c and if one is content with the "coarse grained" time development which gives $|t\rangle$ only for $t = nt_c$. $U[nt_c, (n-1)t_c]$ then is independent of n:

$$U(nt_c, 0) = \exp[-int_c \bar{\mathcal{H}}(t_c)]. \tag{10}$$

The average Hamiltonian need be computed only over one cycle. If one cycle is short: $\|\mathcal{H}\| t_c / \hbar \ll 1$, the series (8), (9), ... will converge rapidly and only the leading terms need be computed.

Note that in principle the form of $\bar{\mathcal{H}}$ may depend on the initial "phase" chosen in defining the cycle of $\mathcal{H}(t)$ Since (10) provides a way of determining the quantum state at the *end* of the n'th cycle, it may be that the behavior of observables $\langle nt_c|Q|nt_c\rangle$ will depend on where in the cycle of periodic behavior of the system the observations are made. As an example, consider the effect of an oscillating r.f. field on an isolated spin:

$$\mathcal{H} = -\omega_0 I_z - 2\omega_1 \cos(\omega t + \phi) I_x \tag{11}$$

or, in the rotating frame

$$\begin{aligned}\mathcal{H}_R &= -(\omega_0 - \omega) I_z - \omega_1 (I_x \cos \phi - I_y \sin \phi) \\ &\quad - \omega_1 [I_x \cos(2\omega t + \phi) + I_y \sin(2\omega t + \phi)].\end{aligned} \tag{12}$$

As an exercise, show that

$$\overline{\mathcal{H}}_R^{(0)} = -(\omega_0 - \omega) I_z - \omega_1 (I_x \cos \phi - I_y \sin \phi), \tag{13}$$

$$\overline{\mathcal{H}}_R^{(1)} = -\frac{\omega_1^2}{4\omega}(1 - 2\cos 2\phi) I_z - \frac{\omega_1(\omega_0 - \omega)}{2\omega}(I_x \cos \phi + I_y \sin \phi). \tag{14}$$

Eq. (14) embodies the Bloch-Siegert shift of the resonant frequency, defined as that value of ω for which the coefficient of I_z in $\overline{\mathcal{H}}_R$ vanishes. Note that the shift depends on the point ϕ in the r.f. cycle at which the magnetization is considered to be measured. An average over phases gives the "ordinary" Bloch-Siegert shift. Such an average is in effect performed by an RF detector.

2. Frames of Reference: Truncation

Note that the preceding calculation was carried out in the rotating frame. Why not in the laboratory frame, where $\mathcal{H}$ also has the required periodic property? An attempt to do so quickly shows that the series (7) does not converge, because $\mathcal{H} t_c$ is not small. The offending part of $\mathcal{H}$ is the large static Zeeman interaction, which disappears in the rotating frame. This illustrates the general principle that we should choose a frame of reference which suppresses very large but simple interactions. By "simple" is meant amenability to exact analysis by elementary means.

The rotating frame provides an important example of the usefulness of choosing a representation which simplifies magnetic resonance problems when a strong Zeeman field is present. In general, given a state vector ψ which satisfies the TDSE with Hamiltonian $\mathcal{H}$

$$\dot{\psi} = -\frac{i}{\hbar} \mathcal{H} \psi \tag{15}$$

one can define a new frame of reference in which the state vector is $\psi_S = S\psi$, S unitary. ψ_S obeys (15), if and only if $\mathcal{H}$ is replaced by

$$\mathcal{H}_S = S\mathcal{H}S^{-1} + i\hbar \dot{S} S^{-1}. \tag{16}$$

The last term takes account of "fictitious" forces, e.g. Coriolis forces, which appear if one frame is accelerated with respect to the other. The rotating frame is entered by

$$\psi_R = R\psi \; ; R = e^{-i\omega t I_z}. \tag{17}$$

If the Hamiltonian included a strong field $H_0\vec{k}$ and an r.f. field, in addition to an internal Hamiltonian $\mathcal{H}_{int}$, one finds (12) above, plus a term $R\mathcal{H}_{int}R^{-1}$. In (12) one ordinarily omits the time-dependent term on the grounds that it constitutes only a weak *zitterbewegung* which causes no secular perturbation of the system. An alternative statement is that this term is entirely off-diagonal when the main Zeeman effect is diagonal, and thus leads only to a second-order correction to the energy. (Unfortunately, the same is true of the second term of (12), which is usually kept, so one must be careful in applying this argument.) Still a third way of justifying this truncation is to note that it corresponds to terminating the $\bar{\mathcal{H}}$ series at $\mathcal{H}^{(0)}$, a procedure which is justified by the fact that $\omega_1/\omega \ll 1$ for $\omega \sim \omega_0$. This is the viewpoint we shall take.

The corresponding truncation of $R\mathcal{H}_{int}R^{-1}$ depends on the form of $\mathcal{H}_{int}$. For the dipolar interaction among like spins

$$\mathcal{H}_{int} = \mathcal{H}_d = \gamma^2 \hbar^2 \sum_{i<j} \sum (I_i \cdot I_j - 3I_i \cdot r_{ij}r_{ij} \cdot I_j)r_{ij}^{-3} \tag{18}$$

one finds

$$\mathcal{H}_{dR} = \gamma^2 \hbar^2 \sum_{i<j} \sum r_{ij}^{-3} P_2(\cos\phi_{ij})(I_i \cdot I_j - 3I_{zi}I_{zj}) + C + D + E + F. \tag{19}$$

The last four terms, not written out, contain explicit time factors $e^{\pm i\omega t}$ or $e^{\pm 2i\omega t}$. When the "size" of $\mathcal{H}_d$, expressible as a local field H_L, is much less than H_0, the oscillating terms can then be omitted. Again, one is truncating the $\bar{\mathcal{H}}$ expansion at $\bar{\mathcal{H}}^{(0)}$. The first term of (19) is usually called the "truncated dipolar Hamiltonian", $\mathcal{H}_d^0$.

The above is an example of the usefulness of choosing a representation which removes or reduces a large $\mathcal{H}$ for which the motion of the system can be calculated explicity. Sometimes the Zeeman effect is not the only such $\mathcal{H}$. Consider the result just obtained:

$$\bar{\mathcal{H}}_R^{(0)} = -(\omega_0 - \omega)I_z - \omega_1 I_x + \sum_{i<j} \sum b_{ij}(I_i \cdot I_j - 3I_{zi}I_{zj}) \tag{20}$$

under conditions where $H_1 = \omega_1/\gamma$ is large. We are free to make a further change of representation as follows:

$$\psi_{TR} = T\psi_R; \quad T = e^{i\xi I_y}; \quad \tan\xi = \frac{\omega_1}{\omega_0 - \omega}. \tag{21}$$

It is left as an exercise to show that

$$\mathcal{H}_{TR} = -\omega_e I_z + \mathcal{H}_d^0 P_2(\cos\xi)$$
$$+ \sum\sum b_{ij}(3\sin\xi\cos\xi)(I_{xi}I_{zj} + I_{zi}I_{xj})$$
$$+ \sum\sum b_{ij}\left(-\frac{3}{2}\sin^2\xi\right)(I_{xi}I_{xj} - I_{yi}I_{yj}) \tag{22}$$

where $\omega_e = \sqrt{\omega_1^2 + (\omega_0 - \omega)^2}$.

If $\omega_e \gg \gamma H_L$, (22) can in turn be truncated. In the language of $\bar{\mathcal{H}}$, one would first transform further into a frame rotating about the z-axis of (22) at frequency ω_e:

$$\psi_{DTR} = D\psi_{TR}; \quad D = e^{-i\omega_e t I_z} \tag{23}$$

whence

$$\mathcal{H}_{DTR} = \mathcal{H}_d^0 P_2(\cos \xi) + \mathcal{H}_{ns}. \tag{24}$$

$\mathcal{H}_{ns}$ consists of terms oscillating at ω_e and $2\omega_e$, which will not contribute to $\overline{\mathcal{H}}_{DTR}^{(0)}$.

Exercise: Compute $\overline{\mathcal{H}}_{DTR}^{(1)}$.

Note that the sequential averaging, first of $\mathcal{H}_R$ and then of $\mathcal{H}_{DTR}$ is not strictly valid in principle. One should properly transform the entire $\mathcal{H}$ into the doubly rotating frame, and then average once. $\overline{\mathcal{H}}_{DTR}^{(1)}$ will then, for example, include terms which are of first order in ω and zero order in ω_e as well as ones which are first order in ω_e and zero order in ω_0. The sequential method will not cause trouble as long as $\omega_e \ll \omega_0$.

Exercise. Consider the Born-Oppenheimer approximation as an example of coherent averaging theory, using a frame of reference in which the electronic part of the Hamiltonian is time-dependent owing to the orbital motion of the electrons about the nuclei.

3. Dipolar Narrowing

According to (24), in a sufficiently strong r.f. field ψ_{DTR} develops according to $\mathcal{H}_d^0 P_2(\cos \xi)$, which can be made to vanish by choosing $\omega_1 = 2^{1/2}(\omega_0 - \omega)$. In the *DTR* frame the system is then oblivious of the dipole-dipole forces which lead to the usual Bloch decay of magnetization in a solid or to the broadening of the steady-state absorption line.

[Recall that the Bloch decay $g(t)$ and the unsaturated steady-state absorption spectrum $f(\omega)$ are related by a Fourier transform.] How is this effect (Lee-Goldburg experiment) related to observations in the laboratory?

1. $\mathcal{H}_{DTR}$ describes the locus of magnetization at integer multiples of a cycle time $t_c = 2\pi/\omega_e$. The actual magnetization will be oscillating with these points as envelope.

2. Because of the transformation T, a magnetization which precesses about the z axis of the *DTR* frame will appear to precess about a tilted axis of the *R* frame.

3. The transformation expresses the fact that a magnetization which is stationary in the *R* frame precesses at a radio frequency in the laboratory. The laboratory oscillation is usually suppressed experimentally by a phase detector, which then measures M_x or M_y with respect to the *R* frame (u and u and v modes).

Thus in a Lee-Goldburg experiment, begun with a magnetization M_0 in the z direction, M_z in the rotating frame will exhibit a (slowly) damped oscillation at ω_e to an asymptote $M_z = M_0/3$. A r.f. phase detector synchronized with the r.f. field will show a sinusoidal oscillation symmetric about, and damping to a value $M_0 \sqrt{2/3}$. A second phase detector in quadrature to the first will show a similar oscillation, shifted $120°$ in phase, symmetric about the damping to zero.

4. Other Internal Hamiltonians

When interactions other than $\mathcal{H}_d$ are present, they can be included in the averaging program outlined above. Consider the chemical (or Knight) shift

$$\mathcal{H}_\sigma = \gamma \sum_i I_j \cdot \sigma_i \cdot H_0. \tag{25}$$

It is easily verified that

$$\overline{\mathcal{H}}^{(0)}_{\sigma R} = \omega_0 \sum_i \sigma_{zzi} \tag{26}$$

where the z axis is that of the laboratory. In the *DTR* frame

$$\overline{\mathcal{H}}^{(0)}_{\sigma DTR} = \overline{\mathcal{H}}^{(0)}_{\sigma R} \cos \xi \tag{27}$$

which exhibits a scaling factor different from that of $\mathcal{H}_d$. Under the Lee-Goldburg conditions, $\tan \xi = \sqrt{2}$, one has

$$\overline{\mathcal{H}}^{(0)}_{\sigma DTR} = \frac{1}{\sqrt{3}} \overline{\mathcal{H}}^{(0)}_{\sigma R} \tag{28}$$

which means that the precessions caused by shifts off resonance will all be slowed by a factor $\sqrt{3}$ by comparison with what would occur in a normal Bloch decay. Thus, while dipolar effects can be suppressed, chemical shifts are reduced but not destroyed.

A general bilinear interaction

$$\overline{\mathcal{H}}_A = \sum_{i<j} \sum I_i \cdot A_{ij} \cdot I_j \tag{29}$$

includes $\mathcal{H}_d$ as a special case for which $\text{Tr}\, A = 0$. If $\text{Tr}\, A_{ij} = 3J_{ij}$, one has from that part

$$\mathcal{H}^0_J = \sum_{i<j} \sum J_{ij} I_i \cdot I_j. \tag{30}$$

Being a scalar, this Hamiltonian is unaffected by any of the rotational transformations discussed above. Thus $\overline{\mathcal{H}}^{(0)}_{JDTR} = \overline{\mathcal{H}}^{(0)}_J$.

5. Decoupling of Unlike Spins

Consider a system of two unlike spins, I and S, coupled by a dipole-dipole interaction which would under ordinary circumstances lead to a doublet absorption spectrum for each. Imagine that spin I is subjected to a strong r.f. field at exact resonance but that S is left undisturbed. The total Hamiltonian

$$\mathcal{H} = -\omega_{0I} I_z - \omega_{0S} S_z - 2\omega_{1I} \cos \omega_{0I} t\, I_x + \mathcal{H}_{IS} \tag{31}$$

is taken into a frame which makes the H field stationary and supresses the I-spin and S-spin Zeeman effects:

$$R = e^{-i\omega_{0I} t I_z} e^{-i\omega_{0S} t S_z} \tag{32}$$

Coherent Averaging and Double Resonance in Solids

whereupon

$$\mathcal{H}_R = -\omega_{1I}I_x + \mathcal{H}_{IS}^0 + \text{oscillating terms}$$

with

$$\mathcal{H}_d^0(IS) = \frac{-2\gamma_I\gamma_S\hbar^2}{r_{IS}^3} P_2(\cos\theta_{IS})I_zS_z = -2bI_zS_z \tag{33}$$

Now if ω_{1I} is large, it is appropriate as earlier to perform a second rotation about the r.f. field:

$$S = e^{-i\omega_{1I}tI_x} \tag{34}$$

so that

$$\mathcal{H}_{SR} - S\mathcal{H}_d^0(IS)S^{-1} - -2bS_z(I_z\cos\omega_{1I}t - I_y\sin\omega_{1I}t). \tag{35}$$

This is purely oscillatory, so its lowest order average vanishes. That is, no effect of the IS coupling on the motion of the spins can occur: they are decoupled.

Exercise: Observe that scalar coupling in place of dipolar coupling leads to the same result: it is in this field that decoupling has been most commonly used, and there the requirement for lowest-order averaging that $\mathcal{H}_{1I} \gg \mathcal{H}_{\text{coupling}}$ is most easily met experimentally.

6. Recoupling of Unlike Spins

Spppose in the previous example strong r.f. fields had been applied to *both* species simultaneously. For generality this time we allow the excitation frequencies ω_I and ω_S to be slightly away from the resonant frequencies ω_{0I} and ω_{0S}. In addition, allow more than one I spin to be present.

We apply the following transformation:

$$U = e^{-i\omega_e It I_z} e^{i\xi_I I_y} e^{-i\omega_I t I_z} e^{-i\omega_e S t S_z} e^{i\xi_S S_y} e^{-i\omega_S t S_z}. \tag{36}$$

The reader can easily verify that the effective average Hamiltonian is

$$\mathcal{H}_{\text{eff}} = \mathcal{H}_{II}^0 P_2(\cos\xi_I) + \mathcal{H}_{IS}^0 \cos\xi_I \cos\xi_S + \mathcal{H}_\pm(t), \tag{37}$$

$$\mathcal{H}_\pm(t) = e^{-i\frac{\Delta t}{2}(I_z - S_z)} \sum_i K_{iS}(I_{+i}S_- + I_{-i}S_+) e^{i\frac{\Delta t}{2}(I_z - S_z)} \tag{38}$$

where $K_{iS} = b_{iS}/2$.

Now suppose that both I and S are at exact resonance ($\xi_I = \xi_S = \pi/2$), and also that the Hartmann-Hahn condition $\omega_{eI} = \omega_{eS}$ is satisfied. Then

$$\mathcal{H}_{\text{eff}} = -\frac{1}{2}\mathcal{H}_{II}^0 + \sum_i K_{iS}(I_{+i}S_- + I_{-i}S_+). \tag{39}$$

The second term expresses a coupling which has been restored, and, moreover, it is a coupling allowing mutual IS flips which are not allowed for unlike spins in H_0 but without r.f. fields. This fact is the basis for the Hartmann-Hahn and other double resonance methods for observing NMR of rare (S) spins in the presence of abundant (I) spins. When the coupling is established the S spin is normally unpolarized (hot) and the I system is polarized (cold). The S spin heats its neighbors (becoming cold itself) and the heat is dissipated into the surroundings by the II spin flips embodied in $\mathcal{H}_{II}^0$.

7. Dipolar Oscillations

A variation of the previous method is possible in which the I-spins are set off resonance by such an amount that $\xi_I = \tan^{-1}\sqrt{2}$, the magic angle. Then the $I-I$ coupling is also removed, and the S spin exchanges energy with its near I neighbors in an oscillatory fashion, without diffusion of this energy among the I spins. Observation of these oscillations permits determination of the constants K_{iS}, i.e., quantitative information concerning the structural environment of the rare spin.

8. Alchemy

All of the previous development concerns manipulation by which the (effective) Hamiltonian of a system is altered at will. Inasmuch as a substance may be said to be *defined* by its Hamiltonian operator, such manipulations may be thought of as constituting the transmutation of one substance into another. There is nothing wrong with this view in principle. Its apparent strangeness derives from the fact that spin systems, because of their simplicity, relative isolation from their surroundings, and slow time scales of dynamical evolution, are susceptible to a variety of *practical* manipulations which are more difficult in the case of more conventional substances.

Macroscopic Dipole Coherence Phenomena

E. L. Hahn*)

Contents

The Semi-Classical Picture 31
The Two-Level System 32
Radiation Damping 35
Damping of Spins in a Cavity 36
Resonance Interaction of Propagating Pulses 39
Self-Induced Transparency and the Photon Echo 41
References 43

The Semi-Classical Picture

A review of early developments in NMR research would give an account of numerous effects involving coherent magnetic field-dipole interactions that play a common role today in the understanding of coherent laser phenomena. When the Bloch and Purcell groups initiated the science of NMR, it was not anticipated then or for a number of years afterward that the dynamics of NMR could be applied to electric dipole resonance transitions. The concepts of spin states dominated the interests of investigators because of the tradition of the the Rabi atomic beam method and because radio frequency techniques were ripe for application. In NMR and EPR research, curiosity was and still is focused upon lineshape and spectral structure analysis, where the spins act as probes of the medium. Because the applied field energy usually far exceeds total spin coupling energies, applied fields are taken as constant, and the role of cavity coupling in the equations is not considered under this restriction. The dynamics of the field intensity variations themselves therefore cannot become evident. This restriction was removed with the advent of maser-laser principles.

Although the gas microwave absorption technique was flourishing, it was not until Dicke [1] (1954) introduced the idea of radiation coherence from an ensemble of phased electric dipole moments that the classical picture was believed to be applicable to dipole radiation from two-level systems other than magnetic spins. The field intensity variations were viewed in terms of spontaneous coherent emission. Dicke pointed out that the coherent superpositions of atomic two-level states attained by cooperation from initial excited states could be prepared just as in pulsed NMR, after which these states would radiate coherently. In 1957 Feynman, Vernon, and Hellwarth [2] showed

---

*) Supported by the National Science Foundation.

that electric dipole transitions in a two-level system could be formulated by equations which are the analogue of the Bloch torque equations for a spin system. Although the discussion in their paper was restricted to a particular maser problem, they utilized the viewpoint very well known as the semi-classical method. In this particular discussion the classical electric field is a self-consistent solution of both the "optical type" Bloch equations, originating from the density matrix, and Maxwell's equations. A number of treatments [3] and reviews [4] have appeared with the express purpose of clarifying and justifying the assumptions of the semiclassical approach. The approach has been rather successful from an experimental point of view in predicting and accounting for a number of pulsed laser phenomena seen in recent years, which is the subject of this brief review, such as photon echoes [5] self-induced-transparency [6], and some properties of mode-locked lasers. [6]

The semi-classical method implies that experimental conditions are restricted to large electric field amplitudes $\vec{E}(\vec{r}, t)$, and also to large collections of N dipoles. The effective momenta of field and particle oscillators is enormous compared to $\hbar$. The subtle aspects of incoherence due to small quantum fluctuations and spontaneous emission are not easily included in the semi-classical method, and at best is done *ad hoc* under special conditions. In general, the semi-classical method has not been satisfactorily doctored to included partial damping due to spontaneous incoherent emission except in the usual phenomenological way of introducing damping terms as in the original Bloch equations.

The overwhelming majority of coherent transient experiments involve a number N of dipoles prepared in phased array which far exceeds $\sqrt{N}$, and therefore the statistics of quantum field fluctuations will not be important or observable. There has been a semantic trend in applying the term "superradiance" to almost every process of coherent radiation from phased dipole arrays, particularly in coherent optics. Since the founding of Maxwell's equations, it has been known that such dipole source terms give rise to the well known reradiated field intensity, emitted in proportion to N^2. It would seem reasonable not to rename an old Maxwell phenomenon, but to apply "superradiance" instead to those special emission processes in which the initial incoherent condition N $\neq$ $\sqrt{N}$ evolves into the coherent emission condition where the intensity is proportional to N^2.

The Two-Level System

The equivalence of the 2 x 2 transformation Pauli matrices for a two-level system of spins (each with spin I = 1/2) to that of electron second-quantization operations makes it possible to draw a one-to-one correspondence between ground and excited states of spin states to those of electron states. A torque equation of the form

$$\frac{d\vec{P}}{dt} = \frac{2p}{\hbar}(\vec{P} \times \vec{E}_{\text{eff}}) = \kappa(\vec{P} \times \vec{E}_{\text{eff}}) \tag{1}$$

results, where p is the dipole matrix element, $\vec{P}$ is the macroscopic polarization, and $\vec{E}_{\text{eff}}$ is the effective electric field. Eq. (1) follows from the density matrix (ρ) method, where the Hamiltonian is written in the dipole approximation as

Macroscopic Dipole Coherence Phenomena

$$\mathcal{H} = \mathcal{H}_0 - \vec{P}_{op} \cdot \vec{E}(z, t);$$
$$\vec{P} = N\mathrm{Tr}\{\rho \vec{P}_{op}\}; \qquad (2)$$

and the energy of the system is

$$W = N\mathrm{Tr}\{\rho \mathcal{H}_0\}. \qquad (3)$$

The ground state energy is

$$W = W_0 = -N\frac{\hbar\omega_0}{2}$$

for N dipoles with resonance transition frequency ω_0. The correspondence between NMR and optical dipole resonance superposition states is sketched as follows:

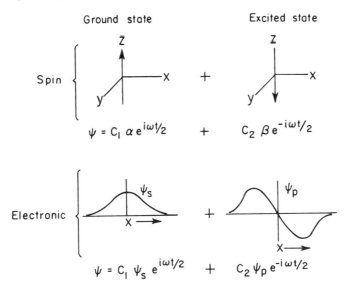

Following a 90° pulse, the superposition state is prepared at $t = 0$ with $C_1 = C_2$, so that a measurement along the x axis will obtain:

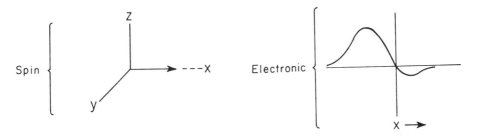

For times $t > 0$ the Larmor phase angle ωt develops, and the precessing spin expectation value appears anywhere in the xy plane. In the electronic case, the electronic wave packet oscillates back and forth between extreme values of $\pm x_{max}$ at frequency ω.

For an assumed system of electronic ψ_s and ψ_p states, the lack of diagonal electronic dipole moments for $t < 0$ in no way detracts from an analogy to the paramagnetic spin case. The magnetization of a spin system is defined as

$$\vec{M} = \hat{u}_0 u + \hat{v}_0 v + \hat{z}_0 M_z \tag{4}$$

in the frame of reference rotating at frequency ω. In the electronic case it is convenient to define

$$\vec{P} = \hat{u}_0 u + \hat{v}_0 v - \hat{z}_0 \frac{W\kappa}{\omega_0} . \tag{5}$$

The quantity $W\kappa/\omega_0$ appears as a fictitious polarization $-P_z$ because of the following correspondence:

$$P_z \equiv M_z = -\frac{W\gamma}{\gamma H_0} = -\frac{W\kappa}{\omega_0}$$

where the usual gyromagnetic ratio is written as $\gamma = \kappa = 2p/\hbar$. Here the energy $W = -M_z H_0$ of the spin system corresponds to the total optical energy $W = N \operatorname{Tr}\{H_0\}$ of the atoms where $|\mathcal{H}_0| = \hbar\omega_0/2$ is determined by the electronic state eigenenergy.

Consider a two-level system tuned at frequency ω_0 off-resonance with respect to an existing electric field plane wave

$$E(z, t) = \mathcal{E}(z, t)\exp\{i[\omega t - kz - \phi(z, t)]\} + \text{cc} \tag{6}$$

propagating in the z direction at carrier frequency ω. Eq. (1) can then be written as

$$\frac{d\vec{P}}{dt} = \vec{P} \times [\hat{u}_0 \kappa \mathcal{E}(z, t) + \hat{z}_0(\Delta\omega + \dot{\phi})], \tag{7}$$

where $\Delta\omega = \omega_0 - \omega$ and $\mathcal{E}$ is a rotating field along the u_0 axis. Using Eq. (5), Eq. (7) may be written out as (with addition of damping terms involving T_1 and T_2)

$$\frac{du}{dt} = v(\Delta\omega + \dot{\phi}) - u/T_2$$

$$\frac{dv}{dt} = -u(\Delta\omega + \dot{\phi}) - \frac{\kappa^2}{\omega_0} \mathcal{E} W - v/T_2 \tag{8}$$

$$\frac{dW}{dt} = v \mathcal{E} \omega_0 - \frac{W - W_0}{T_1} .$$

The Bloch type Eqs. (8) include a possible frequency modulation term $\varphi(z, t)$, which together with $\mathcal{E}(z, t)$ (the slowly varying envelope), are terms to be self-consistent with Eq. (7) and Maxwell's equation:

$$\frac{\partial^2 E(z, t)}{\partial z^2} = \frac{\eta^2}{c^2}\frac{\partial^2 E(z, t)}{\partial t^2} + \frac{4\pi}{c^2}\frac{\partial^2 P(z, t)}{\partial t^2} . \tag{9}$$

In the slowly varying wave approximation,

$$\frac{\partial}{\partial z} \ll \frac{\mathcal{E}}{\lambda} ; \quad \frac{\partial \mathcal{E}}{\partial t} \ll \omega \mathcal{E}$$

is assumed. Combining Eqs. (8) with (9) gives, therefore,

$$\frac{\partial \mathcal{E}(z, t)}{\partial z} = \frac{-2\pi\omega}{\eta c} \int_{-\infty}^{\infty} vg(\Delta\omega)d\Delta\omega \qquad (10)$$

and

$$\mathcal{E}\frac{\partial \phi}{\partial z}(z, t) = \frac{2\pi\omega}{\eta c} \int_{-\infty}^{\infty} ug(\Delta\omega)d\Delta\omega \qquad (11)$$

A retarded time frame of reference is used, where the retarded time is given by $t' = t - \eta z/c$. Eqs. (10) and (11) together with Eq. (8) are useful for the analysis of pulse propagation at or near resonance in a two-level system. These equations can be generalized to cases where the two-level system is degenerate [6, 7], and not necessarily characterized by one dipole matrix element. In this case v and u are replaced by sums $\sum_i v_i$, $\sum_i u_i$ corresponding to sums over i matrix elements.

If the slowly varying wave approximation is not made, and Eq. (9) is applied with no approximations, then terms of the type $\omega \dot{v}$, $\omega \dot{u}$, $\ddot{v}$, $\ddot{u}$ are included also in Eqs. (10) and (11) along with $\omega^2 v$ and $\omega^2 u$ which dominate. It is easily seen [8], however, that the size of these higher order terms is of the same order of magnitude as the off-resonance response contribution by dipole matrix elements in the atom involving transitions to quantum levels other than the two-level system at or on resonance. There is no such thing as a pure two-level optically responsive atom. Other levels must contribute, as is dictated by the sum rule for optical response. Unless other optical levels are taken into account, it is quite possible to arrive at unrealistic conclusions concerning optical pulse propagation effects if Eq. (9) is coupled exactly to Eq. (8), which pertains strictly to a two-level system.

Radiation Damping

The self-consistent field $\vec{\mathcal{E}}$ in the previous discussion is connected with a source term $\vec{P}$ such that stimulated emission and absorption will be naturally accounted for during propagation in the medium, depending upon the initial conditions with regard to $\vec{P}$ and $\vec{\mathcal{E}}$. If $\vec{\mathcal{E}}_i$ is thought of as an initially applied field, then a reaction field $\Delta\vec{\mathcal{E}}$ originating from $\vec{P}$ will add or subtract to $\vec{\mathcal{E}}_i$. The total $\vec{\mathcal{E}}$ however naturally includes any reaction fields, and no extra ones need be added to account for effects due to $\vec{P}$. The self-consistent relationship between $\vec{P}$ and $\vec{\mathcal{E}}$ will automatically include the effects of coherent radiation damping. This is strictly the case for a plane wave propagating through an infinite medium of resonant dipoles. Interference and loss, because of diffraction edge effects due to sample size and finite beam profile, will of course enter the picture if $\vec{P}$ is confined to a finite volume.

If radiation energy is lost from a finite volume element of dipoles, and not recovered by the resonance capture and re-emission of energy in another volume element, it is convenient to introduce what appears to be loss or "coherent radiation damping" terms [1]. Such terms are convenient in cases where the dipole system is first prepared by strong field pulses (considered hardly affected by the medium) in coherent superposition states. After the system is prepared it radiates a coherent field by itself, namely, a spontaneous but coherent "reaction $\vec{\mathscr{E}}$" field. The photon echo is a clear cut example of this [5].

Consider ordinary propagation of a plane wave in the form of a pulse envelope $E(z, t)$ where

$$E(z, t) = \mathscr{E}(z, t) e^{i(\omega t - kz)} + \text{cc.}$$

According to linear absorption behavior, the beam intensity reduction in a medium which contains two-level absorbers is

$$\mathscr{E}^2(z) = \mathscr{E}^2(z=0) e^{-\alpha z}$$

and α is the linear absorption coefficient. Consider a volume element $A\Delta z$ of the medium at position z where A is the area of cross-section. The field amplitude reduction is given by $\Delta\mathscr{E} = -\alpha\Delta z$ after propagation through the slab. Suppose the field source $\mathscr{E}^2(z=0)$ is abruptly cut off. Neglecting the transit time for light to traverse from $z=0$ to $z=z$, there remains a real field $\Delta\mathscr{E}$ which was the reaction field $-\Delta\mathscr{E}$ during the absorption process before $\mathscr{E}$ was cut off. The radiated field $\Delta\mathscr{E}$ has as its source term the polarization v given in Eq. (8), where $v = N_0 p \sin\theta$ and $\theta = \cos^{-1}(v/M_z)$ is the steady state angle at which the total polarization in $A\Delta z$ was tipped when perturbed by $\mathscr{E}$ in the past. In small volume elements therefore, we may imagine the existence of reaction fields $\Delta\mathscr{E}$ such that the resultant field $\mathscr{E} = \mathscr{E}_f = \mathscr{E}_i \pm \Delta\mathscr{E}$ appears for absorption ($-$sign) and for emission ($+$sign). The exit field from the volume element slab is $\mathscr{E}_f$, and the entry field is $\mathscr{E}_i$.

Damping of Spins in a Cavity

In an external magnetic field $\vec{H}_0$, the spin energy of a magnetic moment M_0 tipped from the field $\vec{H}_0$ direction by an angle φ at any time t is given by

$$W_{\text{spin}} = -\xi V_c M_0 H_0 \cos\varphi. \tag{12}$$

The z component of magnetization is $M_z = M_0 \cos\varphi$, ξ is the sample filling factor, and V_c is the cavity volume. With the voltage of nuclear induction given by

$$V = 4\pi N A \xi \omega M_0 \sin\varphi, \tag{13}$$

and $v = M_0 \sin\varphi$, the energy conservation condition requires that

$$\frac{dW_{\text{cavity}}}{dt} = -\frac{dW_{\text{spin}}}{dt} = \frac{V^2}{2R} \tag{14}$$

in an LCR circuit tuned to nuclear resonance at $\gamma H_0 = \omega_0$. After a superposition state has been established by a θ_0 pulse at $t = t_0$, it follows from Eqs. (12), (13), and (14) that the free precession rf field generated by the spins is

$$H_1(t) = \frac{1}{\gamma \tau_r} \text{sech}\,[(t + t_0)/\tau_r] \tag{15}$$

where $\varphi = \gamma H_1(t)\tau_r$. The radiation damping rate is expressed by

$$\frac{1}{\tau_r} = 2\pi M_0 \gamma Q \xi$$

with $Q = \omega L/R$, assuming $Q/\omega \ll \gamma H_1$, and no inhomogeneous broadening is present. For $\theta_0 = 90°$ the free nuclear precession signal power is

$$P_{max} = 8\pi \xi^2 Q M_0^2 \omega V_c \approx \frac{M_0 H_0}{\tau_r} V_s \qquad (16)$$

with $\xi = V_s/V_c$, and V_s is the sample volume.

The coherent damping rate $1/\tau_r$ pertains to coherent radiation into a cavity volume V with a single mode confined to a bandwidth $\Delta \nu = \nu/Q$. The density of states is therefore $\rho_c = 1/(V\Delta\nu)$. With no cavity structure, the dipole system would radiate incoherently because of vacuum field fluctuations H_{vac} at the much slower Einstein spontaneous emission rate:

$$\frac{1}{\tau_E} = \frac{32\pi^3 p^2}{\hbar \lambda^3} = \frac{2\pi}{\hbar^2} \langle p \cdot H_{vac} \rangle^2 \rho_F$$

which involves the free space density of states

$$\rho_F = \frac{8\pi^2 \nu^2}{c^3}.$$

Since NV_s dipoles are radiating in phase in the cavity, then

$$\frac{1}{\tau_r} = \frac{NV_s}{\tau_E} \frac{\rho_c}{\rho_F} = 2\pi M_0 \gamma Q \xi.$$

The coherent emission rate is enhanced by NV_s coherent dipoles and a larger density of states ρ_c.

1. Effect of Inhomogeneous Broadening on the Radiation Damping

The equation for an LCR cavity coupled at resonance with the spins is

$$\frac{d\omega_1}{dt} + \frac{R}{2L}\omega_1 = \frac{-\omega\gamma\pi V_s}{V_c} \int_{-\infty}^{\infty} v(\Delta\omega, t)g(\Delta\omega)d\Delta\omega \qquad (17)$$

where $\omega_1 = \gamma H_1$, and the inhomogeneous field distribution is given by $g(\Delta\omega)$, normalized so that $\int_{-\infty}^{\infty} g(\Delta\omega)d\Delta\omega = 1$. For

$$\frac{R}{L}\omega_1 \gg d\omega_1/dt,$$

Eq. (17) is written as

$$\frac{d\varphi}{dt} = -\frac{1}{M_0 \tau_r} \int_{-\infty}^{\infty} g(\Delta\omega)v(\Delta\omega, t)d\Delta\omega \qquad (18)$$

where $v(0, t) = M_0 \sin \varphi$. If $\varphi = \theta_0$ at $t = t_0$, Eq. (17) yields, after some manipulation [10],

$$\Delta \theta = \phi(\infty) - \theta_0 = -\frac{1}{M_0 \tau_r} v(0, t = \infty) \pi g(0)$$

or

$$\Delta \theta = -\frac{\pi}{\tau_r} g(0) \sin(\theta_0 + \Delta \theta) \tag{19}$$

where $\phi(\infty)$ is the final tipping angle at $t = +\infty$. Suppose $\theta_0 = \pi/2$. Then

$$-\frac{T_2^*}{\tau_r} \cos \Delta \theta \approx -\frac{T_2^*}{\tau_r} = \Delta \theta, \text{ for } \Delta \theta \ll 1 \text{ and } T_2^*/\tau_r \ll 1.$$

The free precession signal lasts for a time T_2^*, during which time the energy $\Delta W = -M_0(\cos \theta(\infty) - \cos \theta_0) H_0 \approx -M_0 H_0 \theta_0 \Delta \theta$ is given up to the cavity as signal energy. The free precession signal power is therefore

$$\frac{dW}{dt} \approx P_{\max} \approx \frac{-M_0 H_0 \theta_0 \Delta \theta}{T_2} \approx \frac{M_0 H_0}{\tau_r} \tag{20}$$

near $t = 0$, just after a $\pi/2$ pulse ($\theta_0 = \pi/2$), the same power of free precession as given by Eq. (16), where inhomogeneous broadening is absent. The result of Eqs. (16) and (20) show as an example that the signal power extracted by the cavity from the coherent system near $t = 0$ is the same whether or not inhomogeneous broadening is present. At any time t, M_0 is reduced by inhomogeneous broadening, but also by damping time constants T_2, T_1 which may be shorter or longer than τ_r. It is this power expressed by Eq. (20) which makes possible the observation of various types of echo and free precession signals.

2. Coupling to the Cavity With No Resistive Losses

For $R/L \to 0$ in Eq. (17), and $T_2^* = \infty$,

$$\frac{d^2\varphi}{dt^2} = N p \pi \gamma \omega_0 \xi \sin \varphi = \frac{1}{\tau_c^2} \sin \varphi.$$

A number of pendulum solutions are assigned to this equation. The finite single pulse solution expresses the spontaneously developed rf field as

$$H_1(t) = \frac{2}{\gamma \tau_c} \sin \varphi/2 = \frac{2}{\gamma \tau_c} \text{sech } t/\tau_c$$

if M_0 is initially inverted at $t = -\infty$ and then tipped infinitesimally to get it to radiate into the cavity. The maximum spin energy is $\Delta W_{\max} = V_s M_0 H_0 (\cos \varphi - 1)$; and the maximum cavity energy is $(2H_1)^2 V_s/8\pi = W_c$. It is seen from above that $W_c/\Delta W_{\max} = 1$. The time constant τ_c is a characteristic coherence or cooperation time serving to define a coherence volume relevant to discussions concerning coherent propagation and radiation [11].

The single mode density of states function given by

$$\rho_c = \frac{1}{V\Delta\nu}$$

in a cavity can be extended to account for resurgence of energy back from the cavity into the spins. For $Q/\omega \ll \gamma H_1$ it was implied previously that the resistance R absorbs all radiated photons from the spins and never returns them to the spins. This is not the case if Q is too high, so that we now have

$$\Delta\nu \cong \frac{\omega}{Q} + \frac{1}{\tau_r} .$$

If $Q \to \infty$, the bandwidth is determined completely by the inverse period of the coherent oscillation frequency for energy exchange between cavity and spins. This same concept applies to field atom energy exchange for propagating optical systems with no losses. Therefore if we express $1/\Delta\nu = \tau_R$ in place of Q/ν, following the discussion after Eq. (16), the expression

$$\frac{1}{\tau_c} = \sqrt{N\rho\pi\gamma\omega_0\xi}$$

is obtained for the coherent radiation time.

For plane wave propagation the density of states is given by $\rho_{PW} = 2\ell/c = 2/\Delta\nu_{PW}$ where $1/\Delta\nu_{PW}$ is a kind of inverse transit time, say for a propagating pulse. The region of pulse excitation is considered to extend over a coherence length ℓ. The distance ℓ cannot be arbitrarily large over which the dipoles will communicate in coherent phase relations.

Resonance Interaction of Propagating Pulses

The self-consistent solution of Eqs. (8) and (9) yields the distance dependence for the pulse area

$$\theta(z) = \kappa \int_{-\infty}^{\infty} \mathcal{E}(z, t)dt$$

to be

$$\frac{d\theta}{dz} = \pm \frac{\alpha}{2} \sin\theta . \tag{21}$$

For an absorber the minus sign applies, and for an amplifier the plus sign applies. Eq. (21) is valid in the slow wave approximation, where frequency modulation is neglected ($\phi = 0$), and the applied frequency ω is set at the center of a symmetric $g(\Delta\omega)$ distribution. The pulsewidth τ may be greater or less than $g(0) \sim T_2^*$. For an absorber all input areas evolve toward zero, 2π, or multiples thereof; and for an amplifier, the areas evolve unstably toward π or odd multiples thereof. The angle measures the quantum mechanical superposition of states at exact resonance ($\Delta\omega = 0$) imposed by the pulse (after $t = +\infty$). The superposition is given by

$$\psi = \psi_1 \cos \pi/2 + i\psi_2 \sin \theta/2.$$

Eq. (21) can be interpreted as expressing the evolution of the Fourier amplitude of the pulse at exact resonance in the interaction representation (or rotating frame). Thus, if we define the Fourier transform as

$$\tilde{\mathscr{E}}(z) = \theta/\kappa = \int_{-\infty}^{\infty} \mathscr{E}(z,t) \exp(i\omega t - \omega_0 t) dt \text{ with } \omega = \omega_0 = 0,$$

a spectrograph will measure the spectral power density function

$$I = \frac{c}{4\pi} \frac{\theta^2}{\kappa^2} = I(0) n^2,$$

where $I(0) = (c/4\pi)\mathscr{E}_{max}^2 \tau^2$, and $\theta = n2\pi$ (for an absorber). For any inhomogeneously broadened two-level system, even with degeneracies as long as level pairs are only excited, one may generally write

$$\frac{d\theta}{dz}(z) \mp \frac{\alpha}{2} S(\theta, z) - \frac{\sigma}{2} \theta$$

and (22)

$$\frac{d\mathscr{I}(z)}{dz} = \mp \alpha(z) F(\theta, z) - \sigma \mathscr{I}(z).$$

The pulse energy is defined by $\mathscr{I} = (\eta c/4\pi) \int_{-\infty}^{\infty} \mathscr{E}(z^2, t) dt$; σ is a scattering loss coefficient, and S and F are nonlinear functions. For a non-degenerate two-level system, $S = \sin \theta$. For a Gaussian pulse with $\tau \gg T_2^*$ it is reasonably (but empirically) accurate to write $F \approx 2(1 - \cos \theta)/\theta^2$. The minus sign applies for an absorber, and the plus sign for an amplifier. In the classical regime $S \approx \theta$, and $F = 1$. In the regime of self-induced transparency, $S = \sin \theta$, $\theta \to \eta 2\pi$, the electric field $\mathscr{E} = 2/\kappa\tau \times \text{sech}\,[(t - z/V)/\tau]$, and $F \to 0$.

In general (for an absorber)

$$\theta(z) = 2 \tan^{-1} \{\exp(-\alpha z/z') \tan \theta_0/2\}$$

and (23)

$$\mathscr{I}(z) \approx \mathscr{I}(0) \exp[-\alpha \int_0^z dz' F(0, z')].$$

The self-induced transparency condition is determined by a process in which energy absorbed from the pulse by the two-level system during the first half of the pulse is returned to the pulse by stimulated coherent emission during the second half of the pulse. The number of photons brought into the volume element of dipoles, in pulse time τ, is approximately equal to the number of dipoles excited if the dipole resonance linewidth is very large compared to the pulse spectral width ($T_2^* \ll \tau$).

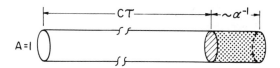

The number of photons is

$$\eta \cong A\frac{\eta\mathscr{E}^2 c\tau}{4\pi\hbar\omega} = AN\frac{T_2^*}{\tau}\alpha^{-1}$$

where T_2^*/τ is the fraction of atoms excited by the Fourier components of the pulse $(1/\tau)$ out of a much broader spectrum $(1/T_2^*)$ of inhomogeneously broadened two-level absorbers. The active dipoles occur over a volume = length α^{-1} × unit area $A(=1)$. Therefore

$$\frac{2p^2}{\hbar^2}\mathscr{E}^2\tau^2 = 1, \text{ or } \theta \approx \kappa\mathscr{E}\tau = \text{constant}.$$

Careful use of the equations gives the constant area condition $\theta = 2\pi$.

Another useful condition which signifies the existence of a stable 2π hyperbolic secant propagating pulse is the fact that it is slowed down to an inverse pulse velocity

$$\frac{1}{V_T} = \eta/c + \frac{1}{V}, \text{ and it is possible to have } V \ll c,$$

so that the medium stores much of the original pulse energy [6]. The expression

$$\frac{d\mathscr{I}}{dz} = -\Delta W \approx \frac{\Delta\mathscr{I}}{\Delta z} = \frac{\eta c}{4\pi}\frac{\mathscr{E}_0^2\tau}{\alpha^{-1}}$$

leads to the ratio [12] of the field energy to the dipole energy in the same volume to be

$$\frac{\eta\mathscr{E}_0^2/4\pi}{\Delta W} = \frac{1}{c\alpha\tau} \sim \frac{V}{c}$$

where self-induced transparency theory gives $1/V \sim \alpha\tau$. For no absorbers present we note that $V = \infty$.

Self-Induced Transparency and the Photon Echo

The photon echo [5] was first obtained and conceived in terms of the response of a small slab of dipole radiators (Cr^{3+} in Al_2O_3) following excitation by a pair of $\mathscr{E}$ field pulses separated by a time T. The superposition of two-level optical states results in a spontaneous echo at $t = 2T$. One argues first that the spontaneous emission rate in erg s^{-1} for a single dipole p_0 is given by

$$\dot{W}_{rad} = \frac{64\pi^4 p_0^2 \eta}{\lambda^3 h}(\hbar\omega). \tag{24}$$

If out of N dipoles per cm^3, an effective macroscopic moment $(Np_0)_{eff}$ is radiating, then $(Np_0)_{eff}$ replaces p_0 in Eq. (24), and $\dot{W}_{rad} \propto N^2$. In carrying out the problem this way, special pains must be taken to introduce a coherence volume ($\sim \lambda^2 \ell$) over which the dipoles cooperate while radiating in the forward direction. The result [5] for the number of photons in the echo is

$$\eta = \frac{\eta c}{4\pi} \frac{\mathcal{E}^2 \tau}{\hbar\omega} = \frac{\ell|\dot{W}_{\text{rad}}|}{\hbar\omega} \propto \frac{N^2 T_2^* \ell^2 p_0^2}{\hbar\lambda\tau} \tag{25}$$

for a sample of unit area cross-section and small thickness $\ell < \alpha^{-1}$.

The area theorem Eq. (21) of self-induced transparency yields the expression for n directly. For two pulses which do not overlap, it has been shown [6] for small sample lengths $\ell = z$ that the echo area is given by

$$\kappa \mathcal{E}_0 \tau = \theta_E \approx \frac{1}{2} \alpha\ell \sin\theta_1(0) [1 - \cos\theta_2(0)], \tag{26}$$

an expression similar to that for magnetic spin-echoes where $\theta_1(0)$ and $\theta_2(0)$ are the input pulse areas at $z = 0$, with pulse width $\tau \ll T$, and T is the pulse separation. The energy in the echo is proportional to θ_E^2/τ, so that

$$\frac{\eta c}{4\pi} \mathcal{E}_0^2 = \frac{\eta c}{4\pi} \frac{\theta_E^2}{\kappa^2 \tau} = W, \text{ and } \theta \approx \alpha\ell.$$

The number of photons in the echo is therefore

$$n \approx \frac{W}{\hbar\omega} = \frac{\eta c}{4\pi} \frac{1}{\kappa^2 \tau} (\alpha\ell)^2 \sim \frac{N^2 T_2^{*2} \ell^2 p_0^2}{\hbar\lambda\tau} \tag{27}$$

which is the same result obtained in Eq. (25). The self-consistent application of Maxwell's equations yields the photon echo in a natural way by taking into account the cooperation volume automatically in the plane wave problem.

Campaan and Abella [13] observed that the intensity I of photon echoes from ruby samples with various concentrations N of Cr^{3+} does not obey the expected coherent radiation dependence $I\alpha N^2$ as N is increased beyond $N \sim 2 \times 10^{-2}$ per cent by weight. They attributed the anomaly to radiation damping corrections, yet to be worked out. However, application of the area theorem as it stands is capable of accounting [14] for their results. As stated before, the photon echo is a natural consequence of the area theorem for an absorber

$$\frac{d\theta}{dz} = -\frac{\alpha}{2} \sin\theta \tag{28}$$

where $\theta = \theta_1(z) + \theta_2(z) + \theta_E(z)$, and α is the Beer absorption coefficient proportional to N. If two pulses with areas $\theta_1(0)$ and $\theta_2(0)$ are applied to the sample at $z = 0$, a third and possibly more pulses with area

$$\theta_E \cong \sum_i \kappa \mathcal{E}_{Ei} \tau_{Ei} \tag{29}$$

will appear subsequently, in order that the total θ can evolve either toward a multiple of 2π or 0. This results because multiple echoes which make up θ_E have phases which add and subtract to $\theta_1(z)$ and $\theta_2(z)$ in such a way as to impose this final condition, $\theta = n2\pi$ at $z = \infty$.

One must be careful to correct Eq. (26), applicable only for $\alpha\ell < 1$, as N increases for a given sample length ℓ to account for the z-dependence of the echo. As the effective sample length is increased, because the number of Beer's lengths $\alpha\ell$ increases, the echo area is modified as a function of pulse propagation distance in the sample, because of the area theorem expressed by Eq. (28). Computer plots show for $\log_{10}\theta_E^2$ plotted versus $\log_{10}\alpha z$ that θ_E^2 reaches a peak and turns over, showing deviation from the $\theta_E^2 \sim N^2$ dependence law. The equations $d\theta_1(z)/dz = -(\alpha/2)\sin\theta_1(z)$ and $d\theta_2/dz = -(\alpha/2)\cos\theta_1(z)\sin\theta_2(z)$ are coupled [7] to Eq. (28). The effective α for the second pulse θ_2 becomes $\alpha\cos\theta$ because [29] of the partial saturation imposed by pulse θ_1, which does not overlap with θ_2. For small echo areas θ_E a reasonably good evaluation of the photon echo is obtained from the equation

$$\frac{d\theta_E}{dz} = \frac{\alpha}{2}\sin\theta_1[1-\cos\theta_2]$$

$$-\frac{\alpha}{2}\sin\theta_E\cos\theta_1\cos\theta_2, \text{ where } \sin\theta_E \sim \theta_E. \tag{30}$$

Therefore for small input areas and small output echo, Eq. (30) yields

$$\theta_E = \frac{\theta_1(0)\theta_2^2(0)}{2} e^{-2q}\sinh q$$

where $q = \alpha z/2$. Note in Eq. (30) that the echo pulse sees an effective absorption coefficient given by $\alpha\cos\theta_1\cos\theta_2$. For small areas θ in general, it is known from previous observations and computer plots that the pulse width τ varies little with z and retains the same order of size as the input τ. Therefore it is acceptable to consider θ_E^2 as proportional to the echo intensity. Also for small $\theta_1(0)$, $\theta_2(0)$, only a single photon echo appears which has a measurable intensity, and subsequent echoes are negligible.

It is significant that the peaking of echo intensity θ_E^2, occurring at $\alpha z = \ln 3$, shows that the α-value at which the turnover occurs ($d\theta_E/dz = 0$) is rather insensitive to various initial values of $\theta_1(0)$, $\theta_2(0)$, for $\theta_1(0) < \pi$. Under these conditions it can be shown that echo T_2' damping does not affect this conclusion.

References

[1] Dicke, R. H.: Phys. Rev. *93*, 99 (1954).
[2] Feynmann, R. P., Vernon, F. L., Jr., Hellwarth, R. W.: J. Appl. Phys. *28*, 49 (1957).
[3] Scully, M., et al.: *Proceedings of the Physics of Quantum Electronics*, Vols. I–IV. Flagstaff, Arizona (1968–1969).
[4] Haken, M.: *Handbuch der Physik*, Vol XXV/2C, Ed. by S. Flügge. Berlin–Heidelberg–New York: Springer 1970. – Allen, L., Eberly, J. H.: *Optical Resonance and Two Level Atoms*. New York: John Wiley & Sons 1975. – Sargent, III, Scully, M. O., Lamb, W. E. Jr.: *Laser Physics*. New York: Addison-Wesley 1974. – Nussenzveig, H. M.: *Introduction to Quantum Optics*. New York: Gordon & Breach 1973.
[5] Abella, I. D., Kurnit, N. A., Hartmann, S. R.: Phys. Rev. *141*, 391 (1966).
[6] McCall, S. L., Hahn, E. L.: Phys. Rev. *183*, 457 (1969).
[7] McCall, S. L., Hahn, E. L.: Phys. Rev. *A2*, 861 (1970).
[8] McCall, S. L., Hahn, E. L.: (to be published).

[9] Bloembergen, N., Pound, R. V.: Phys. Rev. 76, 1059 (1949).
[10] A procedure used to obtain the area theorem is in ref. 6.
[11] Arecchi, F. T., Courtens, E.: Phys. Rev. *A2*, 1730 (1970).
[12] Courtens, E.: Phys. Rev. Letters *21*, 3 (1968).
[13] Compaan, A., Abella, I. D.: Phys. Rev. Letters *27*, 23 (1971).
[14] Hahn, E. L., Shiren, N. S., McCall, S. L.: Phys. Letters *37A*, 265 (1971).

Nuclear Spins and Non Resonant Electromagnetic Phenomena

G. J. Bene

Contents

Low Frequency Range Experiments 45
Excitation in a Rotating Field 46
Experimental Results 48
Excitation in an Alternating Field 50
Bloch-Siegert Effect 50
Non Resonant Dispersion Shift 52
Addendum 54
References 54

A quarter of a century ago, when the magnetic resonance method became the best technique to study nuclear moments, non-resonant effects were neglected mainly because of their small intensity. The following phenomena were observed:

a) The effect of the resonant circular component which rotates opposite to the Larmor precession. The result is a shift of the resonance frequency [1]. Later, the multiple quanta transitions [2] were also observed.

b) The free induction decay following prepolarization performed at a right angle to the precessing field [3]. Today, it is well known that spin echoes can be obtained without resonant pulses and are basically non-resonant phenomena [4].

Included should be also the static magnetization, the absorption and dispersion of relaxation [5], the observation of high order Bloch-Siegert effect using optical pumping, and the observations by C. Cohen-Tannoudji which were interpreted with the concept of the "dressed" atom [6]. These studies were limited to cases where the amplitude and the frequency of the excitation field were larger than the amplitude of the static field and the Larmor frequency of the spin system, respectively.

Low Frequency Range Experiments

It is very convenient to perform experiments on nuclear magnetism of protons in liquid samples in the earth's magnetic field after prepolarization [7]. The sensitivity is strongly enhanced if dynamic polarization is used. In addition, it is very easy to make alternating or rotating fields much larger or much smaller than the earth's magnetic field (0,5 Oe) and the corresponding proton's Larmor frequency. The relaxation times are quite long and it is possible to vary them conveniently.

Excitation in a Rotating Field

First, the simple case of spin $\frac{1}{2}$ in a static magnetic field H_0, directed along the Z-axis and in a rotating magnetic field H_1 of frequency ω in the XY-plane will be considered.

To calculate the interaction, we use the concept of an effective field in the rotating coordinate frame. The rotating frame rotates around the Z-axis with frequency ω. The amplitude of the effective field is given by $H_e^2 = \left(H_0 - \frac{\omega}{\gamma}\right)^2 + H_1^2$. The magnetization of the system of non-interacting spins will precess, in the rotating coordinate frame, around the effective field H_e at the frequency $\omega_e = \gamma H_e$. In the laboratory coordinate frame the frequencies $\omega' = \omega \pm \omega_e$ are observed [8, 9].

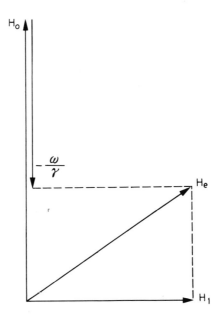

Fig. 1. Effective field in the rotating coordinate frame

A novel approach to the problem has been recently proposed by Fontana [10]. It was proposed to evaluate the frequencies of the observed spectrum and the amplitude of the different components in two extreme cases:

a) The magnetization is along the Z-axis at $t = 0$. Then M_{xy} becomes:

$$M_{xy} \propto \frac{1}{4c^2} \{2\omega_1(c - \Delta\omega)e^{i(\Delta\omega + c)t} - 2\omega_1(c + \Delta\omega)e^{i(\Delta\omega - c)t} + 4\omega_1 \Delta\omega e^{i\Delta\omega t}\} \quad (1)$$

b) The magnetization is in the xy-plane at $t = 0$; the xy-component of the magnetization is given by

$$M_{xy} \propto \frac{1}{4c^2} \{[(c-\Delta\omega)^2 - \omega_1^2]\exp i(\Delta\omega+c)t + [(c+\Delta\omega)^2 - \omega_1^2]\exp i(\Delta\omega-c)t$$
$$+ 2(c^2 + \omega_1^2 - \Delta\omega^2)\exp i\Delta\omega t \qquad (2)$$

where $c^2 = \Delta\omega^2 + \omega_1^2$ and $\Delta\omega = \omega_0 - \omega$.

The system has three characteristic frequencies:

$$\Delta\omega'_a = \Delta\omega + [\Delta\omega^2 + \omega_1^2]^{1/2}, \quad \Delta\omega'_b = \Delta\omega - [\Delta\omega^2 + \omega_1^2]^{1/2}, \quad \Delta\omega'_c = \Delta\omega$$

with $\Delta\omega'_{a,b,c} = \omega_0 - \omega'_{a,b,c}$.

These frequencies are plotted versus $\Delta\omega = \omega_0 - \omega$ in Fig. 2, in which we have on the y-axis the observed displacement $\Delta\omega'$ of the Larmor frequency and on the X-axis the difference $\Delta\omega$ between the Larmor frequency ω_0 and the excitation rotating field frequency ω.

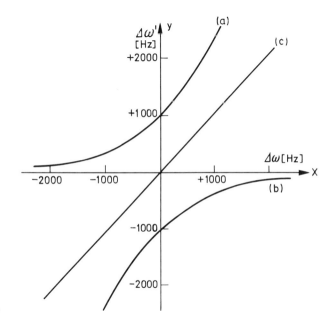

Fig. 2. The three characteristic frequencies of a spin 1/2 system in a rotating field for $H_1 = \omega_1/\gamma$ with $\omega_1 = 1000$ Hz

For an excitation at the Larmor frequency ($X = 0$) we have three frequencies: the maximum of the absorption curve $Y = 0$ and the two maxima of the dispersion curve for $Y = \pm \omega_1$ in the case where ω_1 is large before the natural line width. Amplitudes are given by the coefficients of the exponential functions, but no experimental data are available yet.

Experimental Results

a) On Nuclear Moments

We now summarize the results obtained in a rotating field on nuclear moments under conditions described herein. The resonant frequencies of the nuclei were determined by using a second alternating field to find the natural frequencies of the system $(H_0 + H_1 \cos \omega t)$. Usually, the frequencies are determined directly by observing the damping oscillations. Such determination is done when the magnetization in the initial state makes a large angle with the static magnetic field. The frequency range of the study was $-4000 < \omega < 4000$ Hz, $0 < \omega_1 < 300$ Hz, $\omega_0 = 1971$ Hz. In Fig. 3, the variation of $\Delta\omega_a'$ versus the amplitude of the rotating field at constant frequency $\omega = 800$ Hz is given. In view of the experimental accuracy, the agreement with the theory is excellent.

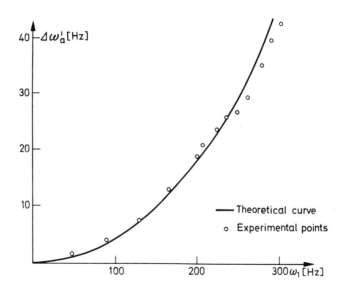

Fig. 3. Displacement of the frequency $\Delta\omega_a'$ versus ω_1 for $\omega = 800$ Hz

The results obtained at constant H_1 and with variable ω are given in Figs. 4 and 5. The experimental points in Fig. 4 are given for ω in the range ± 4000 Hz. The amplitude of the excitation field H_1, in frequency units, is 100 Hz. The theoretical curves for $\Delta\omega_a'$ and $\Delta\omega_b'$ Eq. (3) are also shown. The agreement is excellent.

The details near the resonant frequency are shown in Fig. 5. Far from the resonance frequency only one signal is observed but near the resonance, two signals are detected. The evolution of the signals is in agreement with the Eq. (2). Until now, we have not observed the signal corresponding to the third frequency $\Delta\omega_c'$.

b) In Atomic Magnetism

We are only concerned with the case $\omega_0 \ll \omega, \omega_1$. Transitions ω_a' and ω_b' were studied. If $\omega_0 = 0$, these are $\omega_{a,b}' = -\omega \pm [\omega^2 + \omega_1^2]^{1/2}$. There are three cases: $\omega \gg, =$ or $\ll \omega_1$.

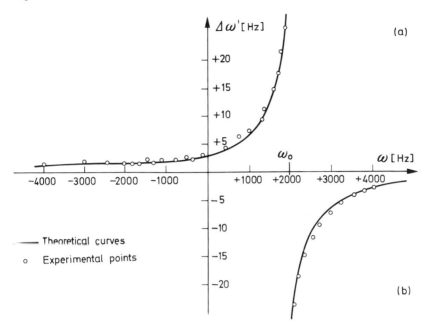

Fig. 4. Displacement of the frequencies $\Delta\omega'_a$ and $\Delta\omega'_b$ versus frequency of the rotating field for $-4000 < \omega < 4000$ Hz and $\omega_1 = 100$ Hz

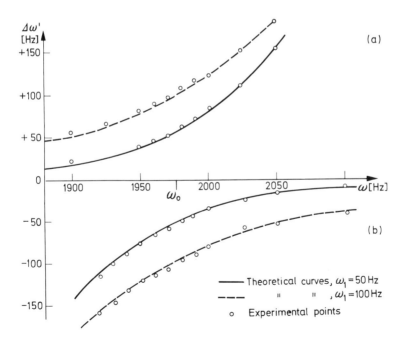

Fig. 5. Displacement of the frequencies $\Delta\omega'_a$ and $\Delta\omega'_b$ versus frequency of the rotating field near the resonant frequency

(1) Condition $\omega \gg \omega_1$, $\omega'_{a,b} \cong -\omega \pm \omega \left[1 - \dfrac{\omega_1^2}{2\omega^2}\right]$. In the second order approximation the two frequencies are $-2\omega - \dfrac{\omega_1^2}{2\omega}$ and $\omega_1^2/2\omega$. Usually, the intensity of the transition at $-2\omega - \dfrac{\omega_1^2}{2\omega}$ is very weak and only the transition at $\omega_1^2/2\omega$ is observed. This frequency corresponds to a fictitious field perpendicular to the plane of the rotating field. M. Le Dourneuf et al. [11] have observed the Larmor frequency of the magnetization for two atoms ^{199}Hg and ^{87}Rb respectively in the states 6^1S_0 and $5^2S_{1/2}$ using an optical pumping technique. The value of the fictitious field H_f is directly related to the amplitude and the frequency of the rotating field, and also to the γ-value of the atomic moments: $H_f = \gamma H_1^2/2\omega$ [11].

(2) and (3) Conditions: no experimental results available.

Excitation in an Alternating Field

It should be noted that the oldest works on electromagnetic interactions in magnetic systems are on non-resonant effects. The works of Gorter and co-workers [5] have been inspired by experiments made by Drude and by theoretical analysis by Debye [12] on phenomena observed when substances having an electric dipole were studied. The RF energy absorption and the variation of the susceptibility in a large frequency range could be interpreted as relaxation effects. The relaxation time, in this case, characterizes the delay between the excitation and the polarization of the system. Far from the resonance frequencies of the system, the dispersion of the susceptibility is given by $\chi' = \chi_0/(1 + \omega^2\tau^2)$ and the energy absorption by the imaginary part of the complex susceptibility $\chi' + i\chi''$ characterizing the excitation by an RF field) $\chi'' = \chi_0\omega\tau/(1 + \omega^2\tau^2)$. Here χ_0 is the static susceptibility, ω the RF frequency and τ the relaxation time of the system χ'' is the "selective" absorption and is different from the resonance absorption. The width of the selective absorption curve at half height is approximately between $\omega = \dfrac{4}{\tau}$ and $\omega = \dfrac{1}{4\tau}$.

Bloch-Siegert Effect

The first non-resonant phenomenon observed and analyzed in nuclear magnetism was the Bloch-Siegert effect [1]. In addition to the main effect, which is a resonance frequency shift, multiple quanta transitions, also shifted from the theoretical frequency, appear. The study of these frequency shifts and multiple quanta transitions in atomic and nuclear systems has been done by optical pumping. In condensed matter, electronic resonance was also used.

A classical calculation gives for the frequency shift

$$\omega_{max(1)} = \omega_0 \{1 + (H_1/2H_0)^2\}^{1/2}$$

and for the three quanta transition frequency

$$\omega_{max(2)} \cong -\omega_0/3.$$

Nuclear Spins and Non Resonant Electromagnetic Phenomena

These results were arrived at by assuming that

$$|\omega + \omega_0| \gg \omega_1.$$

Higher order terms are neglected.

A more complete theoretical approach has been done by Winter [13] and has confirmed the existence of multiple quantum effects in a system with only two atomic levels. Considering the energy and angular momentum conservation laws, it can be seen that several RF photons may be absorbed simultaneously. In our case, a field linearly polarized and perpendicular to Ho is present. It is possible to observe a resonance spectrum at odd harmonics of the frequency of the applied field: $\omega_p = (2p + 1)\omega_0$ where p is an integer. In addition, a Bloch-Siegert frequency shift is observed for all transitions.

For one quantum, the shift equals: $\omega_0 = \omega - (\gamma H_1/4\omega_0)$; for p quanta the relation is

$$\omega_0 = (2p + 1)\omega - \frac{\gamma^2 H_1^2}{4}\left[\frac{1}{p\omega} - \frac{1}{(p+1)\omega}\right].$$

This expression is evaluated to second order.

A higher order analysis of the resonance frequency shifts is necessary to explain the observation of multiple quanta transitions with a strong Bloch-Siegert effect [14, 15]. These phenomena have been observed in sodium vapour [16, 13]. The multiple quanta transitions and their Bloch-Siegert shifts obtained by optical pumping may be seen in Fig. 6. Quantitative values of the H_1 field and amplitudes are not given on this graph.

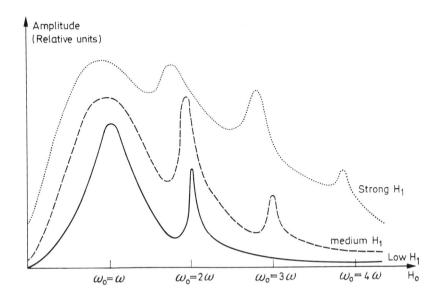

Fig. 6. Multiple quanta transitions and Bloch-Siegert shifts in Na vapour by optical pumping

Non Resonant Dispersion Shift

The Bloch-Siegert effect is a non-resonant phenomenon observed at resonance frequency if the system is excited by an alternating field. The effect, to the lowest order, is a shift of the maximum of the absorption curve and of the symmetry centre of the dispersion curve. When the system is excited far away from the resonance by applying an alternating field H_1 of high amplitude (large compared to the natural width of the signal) no absorption signal has yet been observed. The dispersion curve becomes more and more asymmetric and only the wing near the Larmor frequency is observable over the entire frequency range.

When the system is excited by an alternating field, the difference $\Delta\omega'$ between the frequency of the maximum of the wing ω' and the Larmor frequency ω_0 (no RF present) is not the same as in the rotating field. This is due to the counter rotating component of the alternating field.

Whenever ω and ω_1 are higher than ω_0 the concept of the "dressed atom" has been used. In fact within limits imposed by $\Delta E \cdot \Delta t \leq \hbar$, a quantum system may absorb a nonresonant energy quantum. If the system is in a medium with a high density of photons, it continuously absorbs and re-emits nonresonant photons which form a cloud of "virtual photons". Thus, a "dressed" atom may have different properties than a bare atom.

In atomic physics, the following phenomena were observed (see Figs. 7 and 8).
- The value of the Larmor frequency goes down through zero and becomes negative. It varies as the J_0 Bessel function [17]. Such an evolution is shown in Fig. 7.

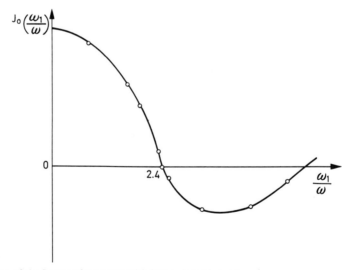

Fig. 7. Variation of the Larmor frequency with increasing values of ω_1/ω

- The Larmor precession is anisotropic; e.g., M_x and M_y are different. The magnetization precesses on an ellipse which becomes more and more flat as the argument of $J_0(\omega_1/\omega)$ goes to the first zero of the function. This appears in Fig. 8 where the precession ellipse is shown for several values of ω_1/ω.

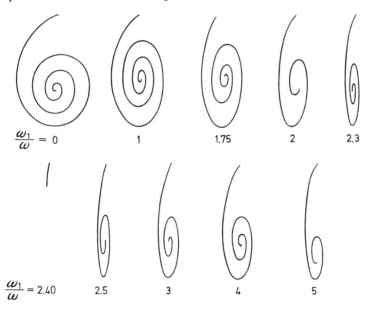

Fig. 8. Anisotropy of the Larmor Precession

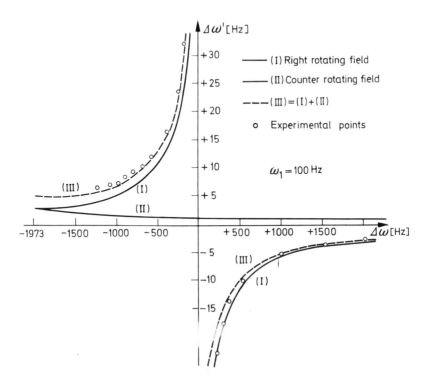

Fig. 9. Non Resonant Dispersion Shift

In nuclear magnetism, no theoretical analysis is available. The first experimental results were obtained in the frequency range between 0 and 4000 Hz at $\omega_0 \cong 2000$ Hz, with $\omega_1 < \omega_0$.

For $\omega_1 = 100$ Hz, the shift $\Delta\omega'$ in the alternating field is approximately the algebraic sum of the shifts of the two rotating components with the same $|\omega|$. The calculated values of these shifts and the experimental points are given in Fig. 9. The non-resonant shift of the dispersion could be very large, because near zero frequency the two rotating components have shifts of the same sign and similar amplitude.

Without giving here any details, it is clear that non-resonant pulses are able to give coherence phenomena or spin echoes. Such works are now in progress in our laboratory.

Addendum

After this lecture was given, some new results were obtained for excitation in a rotating field:

1) The third transition $\Delta\omega'_c$ Eq. (3) was observed in the vicinity of the resonance condition.

2) The observed amplitudes of the three transitions $\Delta\omega'_a$, $\Delta\omega'_b$, $\Delta\omega'_c$ are in good agreement with the theoretical Fontana values, Eq. (2). (January 20th, 1976)

References

[1] Bloch, F., Siegert, A.: Phys. Rev. 57, 522 (1940).
[2] Arimondo, E., Moruzzi, G.: J. Phys. B. Atom. Mol. Phys. 6, 2382 (1973).
[3] Packard, M., Varian, R.: Phys. Rev. 93, 941 (1954).
[4] Bene, G.: Pure Appl. Chem. 32, 72 (1972).
[5] Gorter, C. J.: Paramagnetic Relaxation. Amsterdam: Elsevier Publ. 1947.
[6] Kastler, A.: Pulsed Magnetic and Optical Resonance, p. 1. Ljubljana: Institute J. Stefan 1972.
[7] Performed at the Jussy Station of the DPMC, in collaboration with B. Borcard, E. Hiltbrand and R. Sechehaye
[8] Rabi, I. I.: Phys. Rev. 51, 652 (1937).
[9] Abragam, A.: The Principles of Nuclear Magnetism. Oxford: Clarendon Press 1961.
[10] Fontana, P. R.: Cours 3ème Cycle CICP – Lausanne (to be published) 1975.
[11] Le Dourneuf, M., Cohen-Tannoudji, C., Dupont-Roc, J., Haroche, S.: Compt. Rend. B272, 1048 (1971).
[12] Debye, P.: Polar Molecules, New York: Dover Publ. 1945.
[13] Winter, J. M.: Ann. Phys. (Paris) 4, 745 (1959).
[14] Cohen-Tannoudji, C., Dupont-Roc, J., Fabre, C.: J. Phys. B. Atom. Mol. Phys. 6, L214 and L218 (1973).
[15] Ahmad, F., Bullough, R. K.: J. Phys. B. Atom. Mol. Phys. 7, L147 and L275 (1974).
[16] Margerie, J., Brossel, J.: Compt. Rend 241, 373 (1955).
[17] Haroche, S., Cohen-Tannoudji, C., Audoin, C., Sherman, J. P.: Phys. Rev. Letters 24, 861 (1970)

Nuclear Spin Relaxation in Molecular Hydrogen

F. R. McCourt*)

Contents

The Kinetic Equation and Nuclear Magnetization	57
Correlation Functions for Nuclear Relaxation	58
Spin-Rotation Relaxation	62
Dipole-Dipole Relaxation	63
Application to Hydrogen Gas	65
Acknowledgements	70
References	70

One of the problems of interest to molecular and chemical physicists is the determination of two-body intermolecular potentials. In recent years, with the advent of modern high-speed computers and since the reintroduction of molecular beam techniques, good agreement has been obtained between calculated potentials and those extracted from experimental beam data for noble gas systems. The situation is quite different for the interaction of nonspherical molecules even in the simplest cases. The ordinary bulk transport properties, e.g., diffusion, shear viscosity, thermal conductivity, as well as molecular beam scattering are sensitive primarily to the isotropic part of the intermolecular potential and hence are not suitable for studying anisotropic interactions. There are, however, certain phenomena such as the effect of a magnetic field on transport properties, depolarized Rayleigh light scattering and nuclear magnetic relaxation which depend directly upon the anisotropic parts of the intermolecular potential. It is this relationship for NMR that is to be developed in the present article.

Nuclear magnetic relaxation of the nonequilibrium magnetization ($\vec{M} - M_{eq}$) in liquids and gases was first described phenomenologically by Bloch [1] in 1946. These equations are well known to magnetic resonance devotees and hence need not be repeated here. In liquids and gases, relaxation of the nuclear spin magnetization is caused by random modulation of the nuclear spin alignments by molecular collisions. For monatomic gases, the relaxation is a direct collisional one in the sense that the spin realignment is effected by *inter*-molecular dipole-dipole interactions during a collision or by the interaction of the nuclear spins with the rotational magnetic moment of the collision complex (a diatomic). Such mechanisms are rather weak. Indeed, in noble gases, T_1 is typically of the order of seconds. For polyatomic gases on the other hand, although the effect of these weak intermolecular spin interactions is still

---

*) Alfred P. Sloan Foundation Fellow, 1973–1975.

present, relaxation proceeds via the collisional modulation of the *intra*-molecular nuclear spin-rotational angular momentum and nuclear spin-figure axis interactions since molecular reorientations are caused by ordinary electronic intermolecular interactions, thus completely masking the direct effect of the intermolecular spin interactions. The expectation then, would be that the characteristic relaxation times T_1, T_2 are orders of magnitude shorter than those in noble gases and this is indeed the case. Typical times are shorter than a millisecond.

Even though accurate measurements of T_1 and T_2 as functions of the gas density n (in amagats) over a wide range of n and temperature T are rather difficult to make and are relatively time-consuming, a good deal of progress has been made in this area in recent years as can be seen from the rather comprehensive review articles by Bloom [2] and Armstrong [3]. The task of theory then, is to complete the transition from a knowledge of the single-particle molecular Hamiltonian and, in the dilute gas regime where only binary molecular collisions are of importance, of the form of the anisotropy of the pair potential to formulae for T_1 and T_2 which can be used for the interpretation of the experimental results. This can be achieved by deriving Bloch equations from a microscopic starting point that is concise, simple and which displays the approximations involved.

The transition from a microscopic starting point to expressions for macroscopically measured quantities has traditionally been studied using either general correlation fuction theory or kinetic theory. Perhaps the procedure most widely known to practitioners of NMR is the correlation function method, especially as utilized by Oppenheim und Bloom. [4] Less widely known is the kinetic theory procedure first introduced by Chen and Snider. [5] Correlation function theory offers an elegant general approach to spin relaxation problems but must, in the dilute gas regime, depend ultimately upon a Boltzmann or kinetic equation in order to *calculate* the details of the correlation function as determined by the molecular dynamics of the gaseous system. On the other hand, kinetic theory offers a well-tested procedure for handling the dynamical aspects of gases. However, it is, as traditionally presented [5], less elegant in its treatment of the spin and rotational angular momentum dependence. The theoretical description presented here gives the Bloch equations directly [6] from the kinetic equation by utilizing techniques developed in correlation function theory. In particular, the projection operator formalism introduced by Zwanzig [7] is employed to obtain a "memory-equation" for that part of the distribution function-density matrix characterizing the gaseous system and which is proportional to the nuclear spin operator $\vec{I}$. This equation is then utilized to obtain an integral equation governing the time dependence of the nonequilibrium part of the nuclear magnetization in terms of a kernel having the form of an autocorrelation function.

For the sake of simplicity, only the case of homonuclear diatomic molecules with nuclei having spin-1/2 will be considered here.

The Kinetic Equation and Nuclear Magnetization

The state of the gaseous system is assumed to be described by the single particle Wigner distribution function-density matrix $f(\vec{r}, t; \vec{p}, \vec{J}, \vec{I})$ where $\vec{r}$ and t are position and time and $\vec{p}, \vec{J}$, and $\vec{I}$ are the relevant dynamical variables. The distinction to be made is that the linear momentum $\vec{p}$ is treated classically while the rotational and total nuclear spin angular momenta $\vec{J}$ and $\vec{I}$ are treated quantum mechanically. Further, macroscopic polarizations in the gas are determined as moments of the corresponding single-molecule dynamical operators with f, viz., as

$$n\langle A\rangle(\vec{r}, t) = \int d\vec{p}\; \text{Tr}_Q\, \{A(\vec{p},\vec{J},\vec{I})f\} \equiv \text{Tr}\{A(\vec{p},\vec{J},\vec{I})f\} \tag{1}$$

where Tr_Q designates a trace over the molecular rotational and nuclear spin states of the gas molecules and f is normalized to the absolute equilibrium number density n of the gas. An assembly of gas molecules in thermal equilibrium is governed by the canonical distribution

$$f_{eq}(\vec{p},\vec{J},\vec{I}) = n(2\pi mkT)^{-3/2} Q^{-1} \exp\left\{-\frac{p^2}{2mkT} - \frac{\mathcal{H}}{kT}\right\}. \tag{2}$$

Here Q represents the internal state partition function which is given as $Q \equiv \text{Tr}_Q \exp\{-\mathcal{H}/kT\}$ with $\mathcal{H}$ the internal state Hamiltonian. In the presence of an external magnetic field $\vec{H} = H\hat{h}$, $\mathcal{H}$ is made up of three basic parts corresponding to rotational motion, $\mathcal{H}_{rot}$, Zeeman interactions, $\mathcal{H}_Z$, and intra-molecular couplings, $\mathcal{H}_{intra}$. Thus, for a simple homonuclear diatomic with spin-1/2 nuclei, the internal state Hamiltonian $\mathcal{H}$ has the form $\mathcal{H} = \mathcal{H}_{rot} + \mathcal{H}_Z + \mathcal{H}_{intra}$ in which the rotational contribution is $\mathcal{H}_{rot} = \hbar^2 \vec{J}^2/(2I_0)$ with I_0 the moment of inertia of the molecule. The Zeeman term is $\mathcal{H}_Z = \hbar\omega_I \hat{h}\cdot\vec{I} + \hbar\omega_J \hat{h}\cdot\vec{J}$ and $\omega_I = -\gamma_I H$, $\omega_J = -g_{rot}\mu_N H/\hbar = -\gamma_J H$ are the Larmor precession frequencies associated with the molecule [g_{rot} and μ_N are the rotational g-factor and the nuclear magneton, respectively]. The intra-molecular coupling term has the form

$$\mathcal{H}_{intra} = \hbar\omega_{sr}\vec{I}\cdot\vec{J} + \hbar\omega_d [u]^{(2)} : [\vec{I}_1\vec{I}_2]^{(2)} \tag{3}$$

where ω_{sr} and ω_d are the spin-rotation and dipolar coupling constants, $\vec{u}$ is a unit vector denoting the alignment of the diatomic axis and $\vec{I}_1, \vec{I}_2$ are the individual nuclear spin operators. More will be said about these terms later. An idea of the relative magnitudes of the various parts of $\mathcal{H}$ is given for the H_2 and F_2 molecules in Table 1.

Table 1. Relevant frequencies for NMR in H_2 and F_2. (ω_I and ω_J calculated for a magnetic field of 5 kG)

Molecule	$10^{-8}\nu_{coll}$ (s^{-1}-amagat^{-1})	$10^{-7}\omega_I$[a] (rad-s^{-1})	$10^{-7}\omega_J$ (rad-s^{-1})	$10^{-6}\omega_{sr}$ (rad-s^{-1})	$10^{-6}\omega_d$ (rad-s^{-1})
H_2	3.35[b]	13.4	2.12[c]	.71567[c]	5.4354[c]
F_2	18.5[b]	12.6	−0.29[d]	.9883[d]	0.7545[d]

a Carrington, A., McLaughlin, A. D.: Introduction to Magnetic Resonance. New York: Harper and Row 1967.
b Room temperature value.
c Harrick, N. J., Barnes, R. G., Bray, P. G., Ramsey, N. F.: Phys. Rev. 90, 260 (1953).
d Ozier, I., Crapo, L. M., Cederberg, J. W., Ramsey, N. F.: Phys. Rev. Letters 13, 482 (1964).

Nuclear magnetic relaxation is concerned with the return to equilibrium of the macroscopic polarization of the (dimensionless) nuclear spin vector operator $\vec{I}$, i.e., of the magnetization $\vec{M}(\vec{r},t) \equiv n\gamma_1 \hbar \langle \vec{I} \rangle, (\vec{r},t)$ after an external perturbation has been applied. Since the Zeeman and intramolecular energies are very small compared with kT for all temperatures and all magnetic field strengths normally employed in NMR experiments, f_{eq} can be expanded to terms linear in $(\mathcal{H}_Z + \mathcal{H}_{intra})/kT$ only. This truncation is commonly referred to as the "high temperature" approximation and leads to the Curie law equilibrium polarization which is a rather small quantity. By the same reasoning, since a non-equilibrium magnetization is normally obtained by flipping the equilibrium magnetization with a rf pulse, the resulting system will deviate but little from the equilibrium one and this allows f to be written as

$$f = f_{eq}(1+\Phi) = f_0 \left(1 + \Phi - \frac{\mathcal{H}_Z + \mathcal{H}_{intra}}{kT}\right) \tag{4}$$

where $\Phi = \Phi(\vec{r},t;\vec{p},\vec{J},\vec{I})$ represents the perturbation from equilibrium giving rise to a nonequilibrium system and f_0 is the absolute equilibrium distribution function for molecular translational and rotational motion. It has the same form as (2) but with $\mathcal{H}$ replaced by $\mathcal{H}_{rot}$ and Q by Q_{rot}.

It is clear that once the equation of motion for the nonequilibrium nuclear spin polarization $\langle \vec{I} \rangle^{neq}(t)$ of $\vec{I}$ is known, it can be written in the same form as the Bloch equations to extract expressions for T_1^{-1} and T_2^{-1}. The nonequilibrium polarization $\langle \vec{I} \rangle^{neq}(t)$ is defined through (1) and (4) as $n\langle \vec{I} \rangle^{neq}(t) = n\langle \vec{I} \rangle(t) - n\langle \vec{I} \rangle^{eq} \equiv \langle \vec{I}\Phi(t;\vec{J},\vec{I}) \rangle_0$ where the subscript "o" denotes that a defining relation of the form (1) has been employed but with f_0 replacing f. Notice that reference to the variables $\vec{r}$ and $\vec{p}$ has now been dropped. This corresponds to the assumption of a homogeneous system and to the neglect of all couplings of linear momentum to angular momenta. It is now clear that the time dependence of $\langle \vec{I} \rangle^{neq}(t)$ and hence of $\vec{M}(t) - \vec{M}_{eq}$ can be determined once the time dependence of $\Phi(t)$ is known. The behaviour of $\Phi(t;\vec{J},\vec{I})$ is precisely what the kinetic theory of gases is concerned with and, for a dilute spatially homogeneous polyatomic gas, $\Phi(t)$ is governed by the linearized version of the generalized Boltzmann equation [8]

$$\frac{d\Phi}{dt} = -\mathcal{C}\Phi - \frac{i}{\hbar}[\mathcal{H}_Z,\Phi]_- - \frac{i}{\hbar}[\mathcal{H}_{intra},\Phi]_- \equiv -i\,\mathcal{R}\Phi(t) \tag{5}$$

commonly referred to as the Waldmann-Snider (WS) equation. The commutator operators involving $\mathcal{H}_Z$ and $\mathcal{H}_{intra}$ respresent flow terms in internal state space describing the effect of the externally applied magnetic field and the intra-molecular couplings (internal fields) on the distribution function-density matrix of the gaseous system, is a positive-definite operator representing the effect of binary collisions.

Correlation Functions for Nuclear Relaxation

The Bloom-Oppenheim theory [4] for NMR in a gaseous system requires that certain *ad hoc* assumptions be made regarding the stochastic properties of the system and is not clear in its treatment of the collision dynamics. The Chen-Snider theory [5] treats the collision dynamics in a precise way but, in expanding $\Phi(t)$ in irreducible (Cartesian) tensors in $\vec{J}$ and in $\vec{I}$, can specify easily only the ranks but not the normalizations of the $\vec{J}$-dependent terms. What is desirable is a procedure which combines the elegance of correlation function theory with a precise description of the effect of the molecular collision dynamics.

Only that part $\Psi(t)$ of $\Phi(t)$ which is proportional to $\vec{I}$ is required for the determination of $\langle\vec{I}\rangle^{neq}(t)$ and therefore it is appropriate to introduce a projection operator $\mathcal{P}$ that projects $\Psi(t)$ out of $\Phi(t)$. Of course, $\Psi(t)$ is only of use if an equation governing its time dependence can be obtained from that governing $\Phi(t)$. The essence of the Zwanzig technique is then to employ $\mathcal{P}$ and its complement $\mathcal{Q}$ to do just that. The forms taken by the projection operator and by the expressions for the correlation functions are most clearly expressed in an inner product notation in which the equilibrium average of two quantum mechanical operators A and B is

$$(A, B) \equiv n^{-1} \text{Tr}\{f_0 A^\dagger B\} = n^{-1} \langle A^\dagger B \rangle_0. \tag{6}$$

The dagger represents the quantum mechanical adjoint. In this notation, the requisite projection operator $\mathcal{P}$ is given as

$$\mathcal{P}A \equiv \frac{3}{\vec{I}^2} \vec{I} \cdot (\vec{I}, A). \tag{7}$$

Using $\mathcal{P}$ and $\mathcal{Q}$, $\Phi(t)$ can be split into two parts as $\Phi(t) = \Psi(t) + \chi(t)$, with $\Psi(t) = \mathcal{P}\Phi(t) = 3\vec{I} \cdot \langle\vec{I}\rangle^{neq}(t)/\vec{I}^2$ and $\chi(t) = \mathcal{Q}\Phi(t)$. It is easy to see from these forms that the nonequilibrium polarization of $\vec{I}$ is given as $n\langle\vec{I}\rangle^{neq}(t) = \langle\vec{I}\Phi(t)\rangle_0 = \langle\vec{I}\Psi(t)\rangle_0$ since $\langle\vec{I}\chi(t)\rangle_0$ vanishes identically.

Application of $\mathcal{P}$ and $\mathcal{Q}$ to (5) gives the pair of equations

$$\frac{d\Psi}{dt} = -i\mathcal{P}\mathcal{R}(\Psi + \chi) \; ; \; \frac{d\chi}{dt} = -i\mathcal{Q}\mathcal{R}(\Psi + \chi). \tag{8}$$

If the formal mathematical solution for $\chi(t)$ is substituted into the equation for $\Psi(t)$, the following result is obtained:

$$\frac{d\Psi}{dt} = -i\mathcal{P}\mathcal{R}\Psi(t) - \int_0^t \mathcal{P}\mathcal{R}\exp\{-i(t-s)\mathcal{Q}\mathcal{R}\}\mathcal{Q}\mathcal{R}\Psi(s)ds + \mathcal{P}\mathcal{R}\exp\{-i\mathcal{Q}\mathcal{R}t\}\chi(0). \tag{9}$$

Because of the last term on the right containing $\chi(0)$, this equation is not closed. Under normal experimental conditions, however, $\chi(0)$ vanishes since the deviation from equilibrium is represented initially by a term of the form $\vec{A} \cdot \vec{I}$ in the density matrix produced by disturbing the equilibrium magnetization. However, even if $\chi(0)$ does not vanish identically, the initial value term decreases to zero in a few multiples of the time between two successive reorientation collisions while the times relevant for NMR are normally very much longer. In any case, Eq. (9) is closed if the initial value term is dropped.

The equation of motion for $\langle\vec{I}\rangle^{neq}(t)$ is obtained by taking the moment with $\vec{I}$ of Eq. (9) with $\chi(0) = 0$ as

$$\frac{d\langle \vec{I} \rangle^{neq}}{dt} = -i\frac{3}{\vec{I}^2}(\vec{I}, \mathcal{R}\vec{I}) \cdot \langle \vec{I} \rangle^{neq}(t) - \int_0^t K(t-s) \cdot \langle \vec{I} \rangle^{neq}(s) ds \tag{10}$$

where the kernel $K(t-s)$ is defined by

$$K(\tau) \equiv \frac{3}{\vec{I}^2}(\vec{I}, \mathcal{PR}\exp\{-i\tau \mathcal{Q}\mathcal{R}\}\mathcal{Q}\mathcal{R}\vec{I}). \tag{11}$$

Provided that the direct effect of molecular collisions on $\vec{I}$ can be neglected, the first term on the right-hand-side of (10) is simply the precession term, i.e., $\omega_I \vec{h} \times \langle \vec{I} \rangle^{neq}(t)$. Even with this result substituted into (10), it is difficult to relate the ensuing equation to the Bloch equations largely because a number of simplifications have still to be made in the kernel (11). In particular, $\mathcal{R}$ will now be split into two parts as $\mathcal{R} = \mathcal{R}_0 + \mathcal{R}_1$, where $\mathcal{R}_1$ is due solely to the intramolecular couplings and $\mathcal{R}_0$ contains the effect of collisions and the Zeeman interactions. For the projection operator (7), the two conditions $\mathcal{PR}_1 \mathcal{P} = 0$ and $[\mathcal{P}, \mathcal{R}_0]_{-} = 0$ are satisfied. The first condition is exact and the second holds within the approximation that there is no direct collisional effect on $\vec{I}$. Repeated use of these relations allows (11) to be simplified by straightforward algebraic means to

$$K(\tau) = \frac{3}{\vec{I}^2}(\vec{I}, \mathcal{R}_1 \exp\{-i\tau \mathcal{Q}\mathcal{R}\} \mathcal{R}_1 \vec{I}). \tag{12}$$

Note that $\mathcal{R}$ has been split in such a way that not only do the relations between $\mathcal{R}_0$, $\mathcal{R}_1$ and $\mathcal{P}$ hold but also $\mathcal{R}_1$ is small compared with $\mathcal{R}_0$. Hence, to lowest order in $\mathcal{R}_1$, the operator $\mathcal{R}$ in the exponential in (12) can also be replaced by $\mathcal{R}_0$. With this, the kernel can be reduced to the form

$$K(\tau) = \frac{3}{\vec{I}^2}(\vec{I}, \mathcal{R}_1 \exp\{-i\tau \mathcal{R}_0\}\mathcal{R}_1 \vec{I}) = \frac{3}{\vec{I}^2}(\mathcal{R}_1 \vec{I}, \exp\{-i\tau \mathcal{R}_0\}\mathcal{R}_1 \vec{I}) \tag{13}$$

where, in the latter equality, the fact that $\mathcal{R}_1$ is self-adjoint in the inner product (6) has been used. Finally, (10) can be written as

$$\frac{d\langle \vec{I} \rangle^{neq}}{dt} = \omega_I \vec{h} \times \langle \vec{I} \rangle^{neq}(t) - \int_0^t K(t-s) \cdot \langle \vec{I} \rangle^{neq}(s) ds. \tag{14}$$

It is this result which is to be compared with the Bloch equations.

A major difference between (14) and the Bloch equations is that the coefficients governing the relaxation of $\langle \vec{I} \rangle^{neq}(t)$ as envisaged by Bloch should be time independent while those appearing in the kernel in (14) depend upon time. Of course, Bloch likely had in mind liquids where the characteristic collision frequencies are of the order of $10^{11} s^{-1}$ so that the relevant kernel has associated with it a time scale of $10^{-11} s$, which is very short even when compared with the precessional period of a proton, $\sim 10^{-7} s$ in a 5 kG field. Thus, for liquids a short-memory approximation can be invoked for the kernel, giving directly a Bloch-limit for (14). However, in the gas phase, the situation is not quite so simple since experiments have been carried out [3] at densities ranging from 10^3 to 10^{-2} amagats. Clearly, this corresponds to a range of time scales encompassing the precession period so that the "short-memory" approximation cannot be directly invoked here. Fortunately, the only source of rapid variation in time of $\langle \vec{I} \rangle^{neq}(t)$ is the precession term in (14) and hence it is possible to generate a slowly-varying quantity simply by passing over from the laboratory frame of reference to one which rotates around the static field direction at the nuclear Larmor frequency. This well-known transformation is one which is discussed in every textbook on NMR. The version of the transformation used here is tensorial in nature [6] and relates $\langle\vec{T}\rangle^{neq}(t)$ to its rotating-frame equivalent $\langle\vec{T}\rangle^R(t)$ by $\langle\vec{T}\rangle^{neq}(t) = \exp\{-\omega_I t \vec{h} \cdot \underline{\epsilon}\} \cdot \langle\vec{T}\rangle^R(t)$ where $\underline{\epsilon}$ is the Levi-Civita completely antisymmetrix tensor of rank three (having components $\epsilon_{ijk} = +1$, -1 or 0 depending upon whether ijk is an even, odd or no permutation of xyz). If the relationship between $\langle\vec{T}\rangle^{neq}(t)$ and $\langle\vec{T}\rangle^R(t)$ is used, the governing equation for $\langle\vec{T}\rangle^R(t)$ becomes

$$\frac{d\langle\vec{T}\rangle^R}{dt} = -\int_0^t \mathsf{K}^R(t-s) \cdot \langle\vec{T}\rangle^R(s) ds \tag{15}$$

where the kernel in the rotating frame, $\mathsf{K}^R(\tau)$ is related to $\mathsf{K}(\tau)$ by

$$\mathsf{K}^R(\tau) = \mathsf{K}(\tau) \cdot \exp\{\omega_I \tau \vec{h} \cdot \underline{\epsilon}\} \tag{16}$$

The exponential in (16) can also be represented by $\vec{h}\vec{h} + (\mathsf{U} - \vec{h}\vec{h})\cos\omega_I\tau + \vec{h} \cdot \underline{\epsilon} \sin\omega_I\tau$ in much the same way that the Euler equation relates $\exp\{i\theta\}$ to $\cos\theta$ and $\sin\theta$.

For NMR in a molecular gas, $\mathcal{R}_0$ is made up of three basis parts, viz., $\mathcal{R}_0\Phi = -i\mathcal{E}\Phi + \omega_J[\vec{h} \cdot \vec{J}, \Phi]_- + \omega_I[\vec{h} \cdot \vec{I}, \Phi]_- \equiv (\mathcal{R}_{coll} + \mathcal{R}_Z^J + \mathcal{R}_Z^I)\Phi$. It turns out that the kernel (16) can also be expressed in operator form in terms of an effective rotational Zeeman Hamiltonian operator $\mathcal{R}_Z^{JR} \equiv (\omega_J - \omega_I)[\vec{h} \cdot \vec{J}]_-$ as

$$\mathsf{K}^R(\tau) = \frac{3}{I^2}(\mathcal{R}_1\vec{I}, \exp\{-\tau(\mathcal{E} + i\mathcal{R}_Z^{JR})\}\mathcal{R}_1\vec{I}). \tag{17}$$

Put into words, this is equivalent to saying that the transformation to the rotating frame can be effected by the subtraction of ω_I from all the precession frequencies present. Such a statement has been made long ago [9].

In the rotating frame, the time variation of $\langle\vec{T}\rangle^R(t)$ is caused by the operators appearing in $\mathsf{K}^R(t-s)$ and as can be seen from (17), this is of second order in the intramolecular coupling strengths (see Table I). Thus, on the time scale relevant to the kernel itself, which is of the order of the time between two successive reorientation collisions, $\langle\vec{T}\rangle^R(t)$ is a slowly-varying quantity and a "short-memory" approximation can be invoked to give

$$\frac{d\langle\vec{T}\rangle^R}{dt} \cong -\int_0^\infty \mathsf{K}^R(s) ds \cdot \langle\vec{T}\rangle^R(t). \tag{18}$$

This result has the desired form of the Bloch equations, viz., time-independent relaxation coefficients in the equation of motion for the magnetization,

$$\frac{d\langle\vec{T}\rangle^R}{dt} = -\left[\frac{1}{T_1}\vec{h}\vec{h} + \frac{1}{T_2}(\mathsf{U} - \vec{h}\vec{h}) - \omega_I\sigma\vec{h} \cdot \underline{\epsilon}\right] \cdot \langle\vec{T}\rangle^R(t) \tag{19}$$

with U being the second rank unit tensor, $U_{ij} = \delta_{ij}$. Examination of (19) shows that

T_1^{-1} corresponds to that component of $\langle\vec{I}\rangle^R(t)$ lying along $\vec{h}$, the longitudinal relaxation time, T_2^{-1} corresponds to the transverse relaxation time and, finally, $\omega_{I\sigma}$ corresponds to a true transverse effect, i.e., mutually perpendicular to $\vec{h}$ and to $\langle\vec{I}\rangle^R(t)$ and is a dynamical shift of the position of the resonance line. From a comparison of (18) and (19), T_1^{-1} and T_2^{-1} are given by

$$T_1^{-1} = \vec{h}\vec{h} : \int_0^\infty K^R(s)ds \; ; \qquad T_2^{-1} = 1/2\,(\mathsf{U} - \vec{h}\vec{h}) : \int_0^\infty K^R(s)ds\,. \tag{20}$$

Care has to be taken when evaluating the averages appearing in (17) because the intramolecular coupling commutator operator $\mathcal{R}_1$ introduces correlations between $\vec{I}$ and $\vec{J}$. These correlations have no effect on the way in which $\mathcal{R}_Z^{JR}$ is to be applied, but they do have an effect on the collision term. Later, when considering the specific intramolecular relaxation mechanisms for molecules such as F_2 and H_2, it will be convenient to separate the overall average required by (17) into the product of two individually averaged factors. One factor will involve an average over the rotational states and the other will involve an average over the nuclear spin states. This may be done conveniently by specifying at this stage the effect of the $\mathcal{R}_1$-introduced $\vec{I}$ and $\vec{J}$ correlations. The full collision operator $\mathscr{C}$ can be divided into two parts, one involving the collisional changes for the particle of interest and the other involving collisional changes for the collision partner. Moreover, both terms involve a trace over the nuclear spin states of the collision partner. If, then, direct collisional changes in $\vec{I}$ can be neglected (as is generally the case for all nonmonatomic species) this trace will cause the "partner" terms to vanish (for the same reason that the trace of $\mathcal{H}_Z$ vanishes) leaving only the "self" terms. Thus, a splitting of (17) into tensorial factors depending upon the $\vec{I}$ and $\vec{J}$ parts separately can be effected *provided that* $\mathscr{C}$ in (17) is replaced by $\mathscr{C}'$, the so-called self-only collision operator.

Spin-Rotation Relaxation

Spin-rotation relaxation of the nuclear magnetization in linear molecules is due to the scalar coupling of the nuclear spin of the molecule to the rotational angular momentum. Thus, the fluctuating local magnetic field is caused by the modulation of $\vec{J}$ through reorientation collisions. With the exception of molecular hydrogen, this relaxation mechanism is the predominant one for non-quadrupolar nuclei in diatomic molecules. It predominates, for example [2], for the spin $-1/2$ nuclei in F_2, HF, HD and HCl. Moreover, the spin-rotation relaxation mechanism predominates in the relaxation of spin $-1/2$ nuclei in almost all non-linear polyatomic molecules in the gas phase [2].

For this relaxation mechanism in homonuclear diatomics, $\mathcal{H}_{intra}$ takes the form $\mathcal{H}_{intra} = \hbar\omega_{sr}\vec{I}\cdot\vec{J}$ so that the effect of $\mathcal{R}_1$ on $\vec{I}$ is given as $\mathcal{R}_1\vec{I} = \omega_{sr}[\vec{I}\cdot\vec{J},\vec{I}] = i\omega_{sr}\vec{J}\cdot\vec{\epsilon}\cdot\vec{I}$. Using this result and performing the requisite tensorial manipulations called for by the procedure outlined in the last section, the kernel (17) for the spin-rotation relaxation mechanism becomes

$$K_{sr}^R(\tau) = \frac{1}{3}\,\omega_{sr}^2\langle\vec{J}\cdot,\exp\{-\mathscr{C}'\tau\}\vec{J}\rangle\,\{2\cos(\omega_J - \omega_I)\tau\,\vec{h}\vec{h} + (\mathsf{U} - \vec{h}\vec{h})$$
$$[1 + \cos(\omega_J - \omega_I)\tau] - \vec{h}\cdot\vec{\epsilon}\,\sin(\omega_J - \omega_I)\tau\}. \tag{21}$$

Thus, from (20), the longitudinal relaxation time arising from this mechanism is given explicitly as

$$(T_1^{-1})_{sr} = \frac{2}{3} \omega_{sr}^2 \int_0^\infty (\vec{J}\cdot, \exp\{-\mathscr{C}'t\}\vec{J}) \cos(\omega_J - \omega_I) t \, dt \tag{22}$$

while the corresponding transverse relaxation time is found similarly, from (20) and (21), to be approximated by

$$(T_2^{-1})_{sr} \cong \frac{1}{3} \omega_{sr}^2 \int_0^\infty (\vec{J}\cdot, \exp\{-\mathscr{C}'t\}\vec{J}) [1 + \cos(\omega_J - \omega_I) t] \, dt. \tag{23}$$

A commonly occurring assumption in the literature of NMR in gaseous systems has been that of a single relaxation time for all $\vec{J}$-polarizations of a given rank [4, 5, 9]. One obvious reason for making such an assumption is the attendent exponential decay of the kernel (17) with strong simplification of the formulae for T_1 and T_2. For comparison purposes, it is useful to examine what happens in the present case if such an assumption is introduced. What it consists of then, is the replacement of $\mathscr{C}'$ in the factor $(\vec{J}\cdot, \exp\{-\mathscr{C}'t\}\vec{J})$ appearing in (22) and (23) by τ_{sr}^{-1} such that

$$(\vec{J}\cdot, \exp\{-\mathscr{C}'t\}\vec{J}) = (\vec{J}\cdot, \exp\{-t/\tau_{sr}\}\vec{J}) = \langle \vec{J}^2 \rangle_0 \exp\{-t/\tau_{sr}\}. \tag{24}$$

Of course, this procedure is only valid provided τ_{sr}^{-1} is defined through (24) as

$$\tau_{sr}^{-1} \equiv \langle \vec{J} \cdot \mathscr{C}'(\vec{J}) \rangle_0 / \langle \vec{J}^2 \rangle_0 \tag{25}$$

Using (24) in (22) and (23) gives for T_1^{-1} and T_2^{-1} the following expressions:

$$(T_1^{-1})_{sr} = \frac{2}{3} \omega_{sr}^2 \langle \vec{J}^2 \rangle_0 \tau_{sr} \{1 + (\omega_J - \omega_I)^2 \tau_{sr}^2\}^{-1}$$

and

$$(T_2^{-1})_{sr} \cong \frac{1}{3} \omega_{sr}^2 \langle \vec{J}^2 \rangle_0 \tau_{sr} \{1 + [1 + (\omega_J - \omega_I)^2 \tau_{sr}^2]^{-1}\} \tag{27}$$

which are the standard formulae appearing in the NMR literature. The essential difference between these results and the standard formulae is in the specification of τ_{sr} through the kinetic theory collision integral in Eq. (25).

Dipole-Dipole Relaxation

While the dipole-dipole interaction gives the predominant relaxation mechanism for almost all solids [9], there are very few instances in the gas phase where it plays a significant role. In fact, with the exception of molecular hydrogen, where it is roughly of the same importance as the spin-rotation mechanism, it can normally be neglected as an intramolecular mechanism. The reason is simply that it is only in molecular hydrogen that the distance between the two nuclei is short enough for the direct dipole-dipole interaction to be effective. The tensor operator $[\vec{u}]^{(2)}$ appearing in the second term of Eq. (3) has matrix elements connecting rotational manifolds characterized by the quantum numbers J and $J' = J, J \pm 2$. The off-diagonal matrix elements lead to high-frequency effects in NMR since they will only have an observable effect at densities where the collision frequency matches that characterizing the $\Delta J = \pm 2$ energy differences. For H_2, this would occur at densities in excess of 1000 amagats, well outside the dilute gas regime. Thus, only the matrix elements of $[\vec{u}]^2$ diagonal in J will be retained. These are obtained via the Wigner-Eckart theorem as $[\vec{u}]^{(2)}_{\text{diag}} = -2[\vec{J}]^{(2)}/(4\vec{J}^2 - 3)$, where use has been made of the fact that $\vec{u}$ and $\vec{J}$ are rigorously perpendicular for diamagnetic linear molecules. For the nuclear spin states, only the ortho-species has a nonzero nuclear magnetic moment and $\vec{I} = \vec{I}_1 + \vec{I}_2$ has magnitude 1 so that $[\vec{I}_1 \vec{I}_2]^{(2)}$ can only have matrix elements in the same total I manifold, viz. [5], $[\vec{I}_1 \vec{I}_2]^{(2)}_{\text{diag}} = 1/2[\vec{I}]^{(2)}$. After considerable algebraic and tensorial manipulations, the kernel (17) for the dipolar relaxation mechanism becomes

$$K_d^R(\tau) = \frac{1}{5} ([\vec{\tilde{J}}]^{(2)}:, \exp\{-\mathscr{E}'\tau\} [\vec{\tilde{J}}]^{(2)}) \omega_d^2 \{\vec{h}\vec{h}[\cos(\omega_J - \omega_I)\tau + 4\cos 2(\omega_J - \omega_I)\tau$$
$$+ 1/2 (U - \vec{h}\vec{h})[3 + 5\cos(\omega_J - \omega_I)\tau + 2\cos 2(\omega_J - \omega_I)\tau] -$$
$$- 1/2 \vec{h} \cdot \vec{\epsilon} [\sin(\omega_J - \omega_I)\tau + 2\sin 2(\omega_J - \omega_I)\tau]\} \qquad (28)$$

where $[\vec{\tilde{J}}]^{(2)} \equiv [\vec{J}]^{(2)}/(4\vec{J}^2 - 3)$.

From (20) and (28), the dipolar parts of T_1 and T_2 are obtained as:

$$(T_1^{-1})_d = \frac{1}{5} \omega_d^2 \int_0^\infty ([\vec{\tilde{J}}]^{(2)}:, \exp\{-\mathscr{E}'t\}[\vec{\tilde{J}}]^{(2)}) [\cos(\omega_J - \omega_I)t]dt \qquad (29)$$
$$+ 4\cos 2(\omega_J - \omega_I)t]dt$$

and $$(T_2^{-1})_d \cong \frac{1}{10} \omega_d^2 \int_0^\infty ([\vec{\tilde{J}}]^{(2)}:, \exp\{-\mathscr{E}'t\} [\vec{\tilde{J}}]^{(2)}) [3 + 5\cos(\omega_J - \omega_I)t$$
$$+ 2\cos 2(\omega_J - \omega_I)t]dt \qquad (30)$$

Again, for comparison purposes, it is useful to examine the form taken by (29) and (30) when the single-relaxation-time approximation is introduced. In the case of dipolar relaxation, the $\vec{J}$-polarization is of second rank and a relaxation time τ_d, different from τ_{sr}, has to be introduced such that

$$([\vec{\tilde{J}}]^{(2)}:, \exp\{-\mathscr{E}'t\} [\vec{\tilde{J}}]^{(2)}) = \frac{1}{6} [\vec{J}^2/(4\vec{J}^2 - 3)\rangle_0 \exp\{-t/\tau_d\} \qquad (31)$$

where τ_d^{-1} is defined through (31) to be

$$\tau_d^{-1} \equiv \langle [\vec{\tilde{J}}]^{(2)} : \mathscr{E}'([\vec{\tilde{J}}]^{(2)})\rangle_0 / \langle [\vec{\tilde{J}}]^{(2)}: [\vec{\tilde{J}}]^{(2)}\rangle_0. \qquad (32)$$

Using (31) in (29) and (30) gives for T_1^{-1} and T_2^{-1} in this approximation the expressions

$$(T_1^{-1})_d = \frac{1}{30} \omega_d^2 \langle \vec{J}^2/(4\vec{J}^2 - 3)\rangle_0 \tau_d \{[1 + (\omega_J - \omega_I)^2 \tau_d^2]^{-1}$$
$$+ 4[1 + (\omega_J - \omega_I)^2 \tau_d^2]^{-1}\} \qquad (33)$$

and

$$(T_2^{-1})_d \cong \frac{1}{60} \omega_d^2 \langle \vec{J}^2/(4\vec{J}^2 - 3)\rangle_0 \tau_d \{3 + 5[1 + (\omega_J - \omega_I)^2 \tau_d^2]^{-1} +$$
$$+ 2[1 + 4(\omega_J - \omega_I)^2 \tau_d^2]^{-1}\} \quad (34)$$

which are the standard formulae. As with spin-rotation relaxation, the essential difference between formulae and the standard ones is in the specification of τ_d through kinetic theory in Eq. (32). Notice that in this case the normalization of the second rank $\vec{J}$-tensor polarization is also specified.

Application to Hydrogen Gas

Any theory is tested by its ability to explain as simply as possible the experimental results and by its predictive power. A theory, such as the present one, which is based upon the kinetic theory of gases is valid over a density range which spans the low density and transition regions [3]. Fortunately, this includes much of the available experimental data for simple systems. In particular, hydrogen, which is amenable to detailed collisional calculations, is the only diatomic gas for which extensive and highly accurate measurements are avaible [10] for T_1. In addition, it is the only diatomic gas for which measurements of T_2 have been made as a function of density [11].

Hydrogen has two competitive relaxation mechanisms, the spin-rotation and dipolar intramolecular mechanisms. Their contributions to T_1 and T_2 are additive, since to lowest order in $\mathcal{R}_1$, the exponential operator in (17) preserves the ranks of irreducible tensors in $\vec{I}$ which are produced through the action of $\mathcal{R}_1$ on its right. This is because the collision operator is scalar and the Zeeman operator is rank-preserving. Hence the cross terms between the spin-rotation and dipolar parts of $\mathcal{R}_1$ vanish due to the orthogonality of irreducible $\vec{I}$-tensors. Thus, the results already obtained need simply be added:

$$T_1^{-1} = (T_1^{-1})_{sr} + (T_1^{-1})_d. \quad (35)$$

Before elaborate calculations using the formulae (22) and (29) are undertaken for T_1^{-1}, it is useful to examine the simple single-relaxation-time-approximation (SRTA). This is a modelling procedure which, while it still allows for energetically inelastic collisional contributions, does not allow the rates of relaxation for a given tensor polarziation to vary from one rotational manifold to another: rather, it assigns an average correlation time to the collection of rotational manifolds which are populated at the temperature of the experiment. While such a model makes no assumptions regarding the "strengths" of individual collisions, it does make an assumption about the relevant time scales associated with the energetically inelastic collisional events versus the time scale characterizing the NMR experiment. If the time scale for inelastic collisions is orders of magnitude shorter than the experimentally determined relaxation times (T_1 and T_2), then the experiment should *not* be sensitive to the differences between individual reorientation collisional events but only to some average collision time determined by those events. Even for molecular hydrogen at 10 amagats, the time scale for inelastic collisions is about 1000 times shorter than T_1

From such considerations, SRTA formulae such as (26) and (33) should provide a reasonable description of T_1 as a function of n for many gases. The present procedure, through $\mathcal{R}_1$, generates from $\vec{J}$ the associated $\vec{J}$-dependence. In both cases considered here, only one tensor polarization of each relevant rank is so generated and hence the SRTA model can be expected to be at its best. In such cases the model is tantamount to the assumption that the polarization generated is an approximate eigenoperator of the collision operator with an approximate eigenvalue specified, for example, by (25) or (32).

The correlation times τ_{sr} and τ_d depend upon the effect of collisions upon the quantities $\vec{J}$ and $[\vec{J}]^{(2)}$. Thus, molecular reorientation must occur upon collision if any effect is to be seen in the relaxation experiments. Molecular reorientation cannot be caused by an isotropic interaction potential and so τ_{sr} and τ_d depend directly upon the anisotropic parts of the pair potential. A formal theory for collision matrix elements of the types defining τ_{sr} and τ_d has been recently developed [8] using the distorted wave Born approximation (DWBA) to take into account the anisotropic potential terms. The interaction potential V depends upon the vector $\vec{R} = R\vec{R}$ joining the centres of the colliding pair of molecules and upon the unit vectors $\vec{u}$ and $\vec{v}$ designating the figure axes of the individual molecules. It can be expanded as

$$V(\vec{R}, \vec{u}, \vec{v}) = \sum_{\ell_1 \ell_2 L} V_{\ell_1 \ell_2 L}(R) [\vec{u}]^{(\ell_1)} [\vec{v}]^{(\ell_2)} \cdot {}^{\ell_1 + \ell_2} T(\ell_2 \ell_1 L) \cdot L[\vec{R}]^{(L)} \qquad (36)$$

or, figuratively, as $V(\vec{R}, \vec{u}, \vec{v}) = V_0(R) + \epsilon \, (\vec{R}, \vec{u}, \vec{v})$ where ϵ is an ordering parameter and $V_0 = V_{000}$. In the DWBA procedure, the transition operator t is expanded in powers of the nonspherical interaction as $t = t^{(0)} + \epsilon t^{(1)} + \epsilon^2 t^{(2)} + \ldots$. Terms involving $t^{(0)}$ do not contribute to the collision cross sections important for NMR, so that the leading contribution contains the anisotropic potential terms bilinearly, i.e., $t^{(1)} t^{(1)\dagger}$. The dominant terms in the anisotropic potential for H_2 are the single-P_2 and the quadrupole-quadrupole (QQ) terms corresponding to the $\ell_1 = 2, \ell_2 = 0, L = 2$, the $\ell_1 = 0, \ell_2 = 2, L = 2$ and the $\ell_1 = \ell_2 = 2, L = 4$ terms in (36). Fortunately, there are no cross terms of the form $t^{(1)}_{P_2} t^{(1)\dagger}_{QQ}$ so that the individual contributions of the various anisotropies can be summed.

These calculations are not trivial and the resultant formulae are quite complicated. Nonetheless, if the inverse dependence of a correlation time τ on the number density n is introduced through $\tau = [n \langle v_{rel} \rangle_0 \sigma]^{-1}$ with $\langle v_{rel} \rangle_0 = (16kT/\pi m)^{1/2}$ a mean thermal speed, attention can be focussed upon the corresponding collision cross sections σ_{sr} and σ_d which can be expressed as [12]

$$\sigma_{sr} = A_{202} \sigma^{(0)}_{202} + A_{224} \sigma^{(0)}_{224} \; ; \qquad \sigma_d = B_{202} \sigma^{(0)}_{202} + B_{224} \sigma^{(0)}_{224} \qquad (37)$$

with $A_{202}, B_{202}, A_{224}$ and B_{224} determined as functions of the j quantum numbers by the specific functional forms of (25) and (32). Explicit expressions for these coefficients can be found in Ref. [12]. In principle, of course, the non-negative quantities $\sigma^{(0)}_{202}$ and $\sigma^{(0)}_{224}$ can also be calculated from a knowledge of the R-dependent factors such as appear in (36). This is not easy. In lieu of such *ab initio* calculations, they can be determined by fitting the theoretical result for T_1 [obtained by summing Eqs. (26) and (33)] to the slope of the T_1 vs. n curve in the linear, or intermediate, regime.

Nuclear Spin Relaxation in Molecular Hydrogen

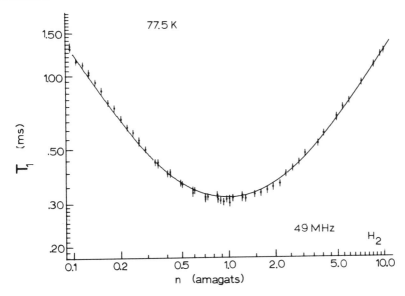

Fig. 1. Comparison of calculated and experimental [10] low density results for T_1 vs. n at a nuclear Larmor frequency of 49 MHz and at 77.5 K in pure H_2. The calculated curve has been fitted to the value of T_1/n in the linear regime. This same comment applies to all of the following figures

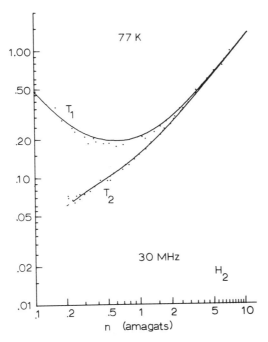

Fig. 2. Comparison of calculated and experimental [10] low density results for T_1 and T_2 vs. n at 30 MHz and at 77.5 K

Molecular hydrogen at 77.5 K represents a true one-level or SRTA system: for this case, all existing theories reduce to the expression first obtained by Schwinger [9]. A one-parameter fit of the 49 MHz data [10] is shown in Fig. 1. As can be seen, the agreement is very good. Also for comparison purposes, the calculated and experimental [10, 11] curves for T_1 and T_2 are given at 30 MHz in Fig. 2.

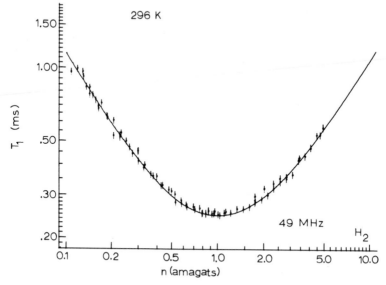

Fig. 3. Comparison of calculated and experimental [10] low density results for T_1 vs. n at 49 MHz and at 296 K in pure H_2

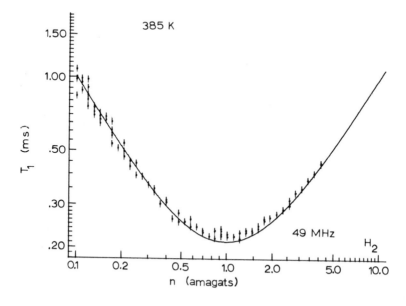

Fig. 4. Comparison of calculated and experimental [10] low density results for T_1 vs. n at 49 MHz and at 385 K in pure H_2

A more detailed test of the SRTA formulae for T_1 in H_2 can only be made at higher temperatures where more than one rotational level is significantly populated. At 296 K, using (37) with calculated coefficients $A_{202} = 0.0583$, $B_{202} = 0.0567$, $A_{224} = 0.000896$ and $B_{224} = 0.000790$, the requirement that σ_{sr} and σ_d fit the value $T_1/n = 0.106$ ms amagat^{-1} as given [10] at 49 MHz places strong bounds upon the allowed values for them: 0.450 Å$^2 < \sigma_d < 0.466$ Å^2 and 0.511 Å$^2 > \sigma_{sr} > 0.479$ Å^2. For $\sigma_d = 0.457$ Å^2 and $\sigma_{sr} = 0.493$ Å^2, $\sigma_{202}^{(0)}$ and $\sigma_{224}^{(0)}$ have the values 4.20 Å^2 and 278 Å^2, respectively. The behaviour of T_1 vs. n at 296 K is shown in Fig. 3. Again, the agreement with experiment is quite good. Fig. 4 shows a similar comparison of T_1 vs. n for 49 MHz at 385 K [10]. The one-parameter fit is still quite satisfactory. In this case, the bounds placed upon σ_{sr} and σ_d by requiring them to fit $T_1/n = 0.098$ ms amagat^{-1} are: 0.43 Å$^2 < \sigma_d < 0.46$ Å^2 and 0.40 Å$^2 > \sigma_{sr} > 0.37$ Å^2. At 385 K, the A and B coefficients are: $A_{202} = 0.04369$, $B_{202} = 0.05477$, $A_{224} = 0.00077$ and $B_{224} = 0.00084$.

Table 2. Collision cross section contributions

			σ_{sr} (Å^2)		
Temperature	Contribution	Elastic	Resonant	Inelastic	Total
77.5 K	P_2	1.2024	–	–	1.2024
	QQ	0.2076	–	–	0.2076
	Total	1.4100	–	–	1.4100
296 K	P_2	0.2441	–	0.0003	0.2447
	QQ	0.1967	0.0392	0.0126	0.2485
	Total	0.4411	0.0392	0.0129	0.4932
385 K	P_2	0.1863	–	0.0006	0.1869
	QQ	0.1364	0.0430	0.0185	0.1980
	Total	0.3227	0.0430	0.0191	0.3848

			σ_d (Å^2)		
Temperature	Contribution	Elastic	Resonant	Inelastic	Total
77.5 K	P_2	0.7216	–	–	0.7216
	QQ	0.1246	–	–	0.1246
	Total	0.8462	–	–	0.8462
296 K	P_2	0.2376	–	0.0002	0.2378
	QQ	0.1913	0.0192	0.0089	0.2193
	Total	0.4289	0.0192	0.0090	0.4570
385 K	P_2	0.2338	–	0.0039	0.2342
	QQ	0.1713	0.0280	0.0165	0.2158
	Total	0.4051	0.0280	0.0169	0.4500

The various contributions to the collision cross sections σ_{sr} and σ_d are summarized in Table II. The contributions fall into three categories: elastic or pure reorientation (for which $\Delta M_J \neq 0$, $\Delta J = 0$, $\Delta\epsilon_{rot} = 0$), resonant (for which $\Delta M_J \neq 0$, $\Delta J \neq 0$ but $\Delta\epsilon_{rot} = 0$) and inelastic (for which $\Delta M_J \neq 0$, $\Delta J \neq 0$ and $\Delta\epsilon_{rot} \neq 0$) collisions for both the P_2 and QQ interactions. In proceeding from 77.5 K to 385 K, there is an increase in the relative importance of the resonant and inelastic contributions, reaching 16 % of σ_{sr} and 10 % of σ_d at 385 K. Nonetheless, at the temperatures shown, these collisions always remain relatively elastic.

This article has attempted to illustrate how the use of kinetic theory together with correlation function techniques allows a systematic derivation of the Bloch equations for a gaseous system and facilitates their use in analyzing experimental T_1 relaxation data. There are numerous approximations involved in all treatments of this problem and one of the objectives here has been to point these out as clearly as possible while at the same time illustrating the amount of meaningful information regarding collision cross sections that can still be extracted from T_1 data.

Acknowledgements

To my graduate student, Mr. T. E. Raidy, I am especially grateful for his calculation of the data appearing in Table II and for his preparation of the figures. To Professor R. L. Armstrong and Dr. K. E. Kisman, I am grateful for supplying us with their raw data points.

References

[1] Bloch, F.: Phys. Rev. 70, 460 (1946).
[2] Bloom, M.: MTP Int. Rev. Sci., Phys. Chem. Ser. 1, 4, 1 (1972).
[3] Armstrong, R. L.: This volume (1976).
[4] Bloom, M., Oppenheim, I.: Advan. Chem. Phys. 12, 549 (1967).
[5] Chen, F. M., Snider, R. F.: J. Chem. Phys. 48, 3185 (1968).
[6] McCourt, F. R., Raidy, T. E., Festa, R., Levi, A. C.: Can. J. Phys. 53, 2449 (1975).
[7] Zwanzig, R.: Boulder Lectures in Theor. Phys. III (Interscience, 1961), 106.
[8] Moraal, H.: Phys. Reps. 17C, 225 (1975).
[9] Abragam, A.: Nuclear Magnetism, Chap. 8. Oxford: University Press, 1961.
[10] Kisman, K. E., Armstrong, R. L.: Can. J. Phys. 52, 1555 (1974).
[11] Hardy, W. H.: Can. J. Phys. 44, 265 (1966).
[12] McCourt, F. R., Raidy, T. E., Rudensky, T., Levi, A. C.: Can. J. Phys. 53, 2463 (1975).

Longitudinal Nuclear Spin Relaxation Time Measurements in Molecular Gases

R.L. Armstrong

Contents

1. General Considerations 71
   1.1 Introduction 71
   1.2 Electric Dipole Spectroscopy 71
   1.3 Nuclear Spin Relaxation Spectroscopy 73
   1.4 Nuclear Spin Symmetry Species 76
2. Low Density Regime 77
   2.1 The Region of the Characteristic T_1 Minima 77
   2.2 A Fundamental Test of Relaxation Theory 77
   2.3 Free Molecule Spin-Rotation Parameters 80
   2.4 The Effects of Centrifugal Distortion 82
3. Intermediate Density Regime 84
   3.1 The Study of Intermolecular Forces by Means of T_1 Measurements 84
   3.2 A Simple Kinetic Theory Approach 84
   3.3 A Semiclassical Scattering Approach to the Calculation of Cross Sections 86
   3.4 The Special Case of Hydrogen Gas 87
4. High Density Regime 90
   4.1 The Passage from Intermediate to High Densities 90
   4.2 A Test of the Enskog Theory 91
   4.3 The Effect of Overlapping Rotational Levels 94
References 95

1. General Considerations

1.1 Introduction

For molecular physicists and physical chemists nuclear magnetic resonance is but one of many experimental methods that can be applied to the study of molecular motions and intermolecular forces in molecular gases [1]. It is instructive to consider the relation between nuclear spin relaxation spectroscopy and electric dipole spectroscopy [2].

1.2 Electric Dipole Spectroscopy

When a system of molecules is subjected to a time dependent electric field $E_x \cos \omega t$ the coupling between the electric dipole moment of the system and the electric field causes transitions between the eigenstates of the system. The probability for transitions between two states of the system is proportional to the square of the matrix element of the dipole moment operator between these states. For weak electric fields the response of the system is linear in E_x and the induced polarization P_x is

$$P_x = \chi'(\omega) E_x \cos \omega t + \chi''(\omega) E_x \sin \omega t$$

where $\chi'(\omega)$ and $\chi''(\omega)$ are the in-phase and out-of phase components of the electric susceptibility. The power absorbed from the electric field is proportional to $\chi''(\omega)$. In a low density gas, collisions are relatively infrequent and $\chi''(\omega)$ is proportional to the Fourier transform of the correlation function G(t) of the single molecule electric dipole moment operator $\mu_x(t)$. That is,

$$\chi''(\omega) \propto \int_{-\infty}^{\infty} \text{Real} \langle \mu_x(0) \mu_x(t) \rangle \exp(-i\omega t) dt = \int_{-\infty}^{\infty} G(t) \exp(-i\omega t) dt$$

where the brackets $\langle \rangle$ denote an equilibrium ensemble average. The correlalation function may be written

$$G(t) = [G(t)]_{\text{free}} \, g(t)$$

where $[G(t)]_{\text{free}}$ is the free molecule correlation function and g(t) the reduced correlation function which describes the effect of the collisions. The function $[G(t)]_{\text{free}}$ is given by

$$[G(t)]_{\text{free}} = \text{Real} \sum_m P_m \langle m | \mu_x(0) \mu_x(t) | m \rangle = \sum_{mn} P_m | \langle m | \mu_x(0) | n \rangle |^2 \cos \omega_{mn} t$$

where P_m is the Boltzmann factor and $\omega_{mn} = (E_m - E_n)/\hbar$ with E_m and E_n the energies of eigenstates $|m\rangle$ and $|n\rangle$. The function $g(t)$ is often approximated by

$$g(t) = \exp(-|t|/\tau_c)$$

where τ_c is a correlation time associated with the collisions. For $t = 0$, $g(0) = 1$ and for $t = \infty$, $g(\infty) = 0$ as required. The corresponding spectral density function $J(\omega)$, which determines the electric dipole spectrum, is

$$J(\omega) = \int_{-\infty}^{\infty} G(t) \exp(-i\omega t) dt = \frac{1}{2} \sum_{mn} P_m | \langle m | \mu_x(0) | n \rangle |^2 [j(\omega - \omega_{mn}) + j(\omega + \omega_{mn})]$$

with

$$j(\omega - \omega_{mn}) = \int_{-\infty}^{\infty} \exp[-i(\omega - \omega_{mn})t] g(t) dt = \frac{2\tau_c}{1 + (\omega - \omega_{mn})^2 \tau_c^2}.$$

The spectrum consists of a set of absorption lines of Lorentzian shape centered at frequencies ω_{mn} and of intensities $P_m |\langle m | \mu_x(0) | n \rangle|^2$. As an example, consider the pure rotational absorption spectrum for a diatomic molecule possessing a permanent electric dipole moment. The relevant eigenstates are characterized by the rotational angular momentum quantum number J and the only non-zero matrix elements $\langle J | \mu | J' \rangle$ of the electric dipole moment operator correspond to $\Delta J = \pm 1$. Fig. 1 shows the rotational electric dipole spectrum for a dilute gas of HD molecules at 300 K.

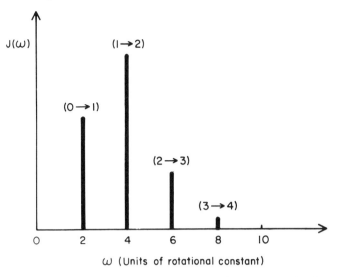

Fig. 1. Schematic illustration of the positive frequency part of the pure rotational spectral densities of a dilute gas of diatomic molecules. The electric dipole spectrum ($\Delta J = 1$) appropriate to HD gas at 300 K is shown. The rotational constant in units cm^{-1} is defined as $\hbar/2cI$ where c is the speed of light and I the molecular moment of inertia

The phenomenon of non-resonant absorption or dielectric relaxation may be thought of as the special case of electric dipole molecular spectroscopy associated with collisional broadening of the $\omega_{mn} = 0$ terms. Non-resonant absorption may be measured at microwave or radio frequencies. For diatomic molecules there are no non-zero matrix elements at zero frequencey. However, for symmetric top molecules, as second quantum number K is required for a complete description of the eigenstates in addition to the elements at zero frequency. However, for symmetric top molecules, a second quantum number K is required for a complete description of the eigenstates in addition to the quantum number J. K is a measure of the projection of $\vec{J}$ along the figure axis of the molecule. The matrix elements $\langle JK|\mu|J'K'\rangle$ are non-zero for $\Delta J = 0, \pm 1$ and $\Delta K = 0$ if the permanent dipole moment lies in the direction of the figure axis. Non-resonant absorption occurs for such molecules. It is relatively easy to measure the dispersion term $\chi'(\omega)$ at low frequencies. The Kramers-Kronig relations then permit a deduction of the absorption term $\chi''(\omega)$.

1.3 Nuclear Spin Relaxation Spectroscopy

In many cases the approach of the nuclear magnetization associated with a molecular gas to its equilibrium value M_0, is well approximated by an exponential function of time. In the presence of a static external magnetic field in the z-direction

$$M_z(t) = M_0 - [M_0 - M_z(0)] \exp(-t/T_1)$$

where T_1^{-1} is the longitudinal relaxation rate. It is a measure of the rate at which energy is exchanged between the nuclear spin system and the other degrees of freedom called the lattice. For a dilute molecular gas the spin-lattice coupling usually results from intramolecular spin-dependent interactions such as the spin-rotation, magnetic dipolar and quadrupolar interactions. The rate constant T_1^{-1} is governed by the spectral densities of the correlation functions of the relevant lattice variables for the spin-lattice interactions; the spectral density is evaluated at frequencies corresponding to the energy differences between nuclear spin states coupled by the spin-lattice interactions. As a result of collisions within the gas the spectral density of the intramolecular interactions is distributed over a range of frequencies about the pure rotational frequencies. The frequency range is of order τ_c^{-1} where τ_c is the correlation time for collisions which interrupt the molecular rotation or which cause molecular reorientation. The precise relation between T_1^{-1} and the spectral density has been developed quite rigorously in the general theory of magnetic resonance [3]. It follows that measurements of T_1^{-1} as a function of the nuclear Larmor frequency ω_0 should give a maximum each time that ω_0 is tuned to a rotational frequency associated with a pair of molecular levels between which the spin-lattice interactions have non-zero matrix elements. Such a nuclear spin relaxation spectroscopy experiment is analogous to the usual experiment in electric dipole molecular spectroscopy. The selection rules for a diatomic molecule are $\Delta J = 0$ for the spin-rotation interaction and $\Delta J = 0, \pm 2$ for the dipolar and quadrupolar interactions. Fig. 2 shows the proton spin relaxation spectrum for a dilute gas of HD molecules at 300 K. However, since the largest Larmor frequency possible corresponds to a few hundred MHz, whereas typical rotational frequencies lie in the infrared, no nuclear spin relaxation spectroscopy experiment has ever been carried out.

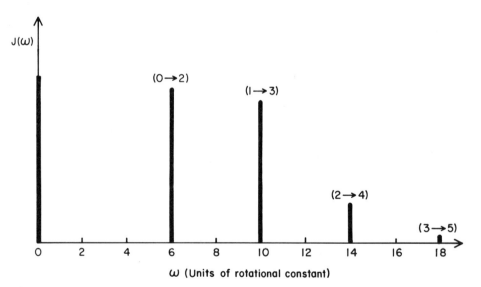

Fig. 2. Schematic illustration of the positive frequency part of the pure rotational spectral densities of a dilute gas of diatomic molecules. The nuclear spin relaxation spectrum ($\Delta J = 0,2$) appropriate to HD gas at 300 K is shown

In practice, measurements of T_1^{-1} are analogous to non-resonant absorption measurements and T_1^{-1} is determined primarily by that part of the spin-lattice interaction which couples free molecular states of the same energy. The contribution to

T_1^{-1} of matrix elements of the spin-lattice interaction between rotational states of different energy is quenched. For diatomic gases at high temperatures the contributions of the dipolar and quadrupolar interactions are reduced to 1/4 of their total values, whereas there is no quenching for the spin-rotation interaction since only $\Delta J = 0$ transitions are allowed. In general, the lattice operators associated with the dipolar and quadrupolar interactions tend to be quenched much more in the gas than those associated with the spin-rotation interaction. The opposite is true in solids. The reason for this difference is that J is a good quantum number for free molecules while in a solid, crystalline electric fields of low symmetry remove the rotational degeneracy thereby quenching J.

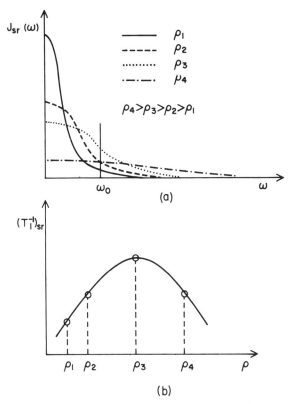

Fig. 3. (a) Illustration of the spectral density $J_{sr}(\omega)$ of the spin-rotation interaction at several different densities. (b) Schematic illustration of the relaxation rate $(T_1^{-1})_{sr}$ in a dilute molecular gas as a function of density. The relative relaxation rates associated with the spectral densities in (a) are shown

As an example, consider the nuclear spin relaxation due to the spin-rotation interaction in a dilute gas of diatomic molecules. The intramolecular coupling is of the form

$$\mathcal{H}_{sr} = -\hbar C_{sr} \vec{I} \cdot \vec{J}$$

where $\vec{I}$ is the nuclear spin angular moment, $\vec{J}$ the molecular rotational angular momentum and C_{sr} the scalar coupling constant. The free molecule correlation function is

$$[G(t)]_{\text{free}} = \frac{C_{sr}^2}{3} \frac{2I_0 kT}{\hbar^2}$$

where I_0 is the molecular moment of inertia. In this case $[G(t)]_{\text{free}}$ is a constant as it must be since the angular momentum of a freely rotating molecule is a constant of the motion. The associated spectral density function is

$$J_{sr}(\omega) = \frac{4}{3} \frac{I_0 kT}{\hbar^2} C_{sr}^2 \frac{\tau_{sr}}{1 + \omega^2 \tau_{sr}^2}$$

where the correlation time τ_{sr} is inversely proportional to the gas density ρ. Fig. 3(a) shows $J_{sr}(\omega)$ as a function of ω for several values of ρ. The relation between the relaxation rate $(T_1^{-1})_{sr}$ and $J_{sr}(\omega)$ is

$$(T_1^{-1})_{sr} = 4\pi^2 J_{sr}(\omega_0).$$

Therefore, $(T_1^{-1})_{sr}$ will exhibit a maximum for the gas density for which $\omega_0 \tau_{sr} = 1$. Fig. 3(b) shows the variation of $(T_1^{-1})_{sr}$ with density.

1.4 Nuclear Spin Symmetry Species

For a homonuclear diatomic molecule in the ground electronic state the operation of permuting the nuclei is formally equivalent to the parity operation. Because all the molecular hyperfine interactions are invariant with respect to the parity operation, there is no mixing of nuclear spin states which are even or odd with respect to permutation. The total molecular wave function is even for nuclei like deuterons that obey Bose-Einstein statistics and odd for nuclei like protons that obey Fermi-Dirac statistics. It follows that the quantum states of a homonuclear diatomic molecule may be rigorously classified into two 'nuclear spin symmetry species'. The species with the greater nuclear spin statistical weight is called ortho; the other species is called para. Conversion between the two species is known to be extremely slow for hydrogen and deuterium and is presumed to also be slow for other homonuclear diatomic gases. For many practical purposes a gas of homonuclear diatomic molecules may be thought of as consisting of two non-interacting component gases. One might therefore anticipate that the ortho and para species will relax at different rates so that the relaxation curve will contain the sum of two exponential relaxation terms. For nuclei having spin $I > 1/2$ the dominant relaxation mechanism is provided by the quadrupolar coupling to the fluctuating local electric field gradients. The magnetization of the ortho and para species each relax exponentially with a rate

$$(T_1^{-1})_q^i = \frac{3\pi}{20} \left(\frac{eqQ}{\hbar}\right)^2 [J_q^i(\omega_0) + 4J_q^i(2\omega_0)]$$

with the spectral density function $J_q^o(\omega)$ evaluated for the ortho species using an ensemble of molecules characterized by even parity vibration-rotation eigenfunctions and with the function $J_q^p(\omega)$ evaluated for the para species using odd parity vibration-rotation eigenfunctions. The quantity ($eqQ/\hbar$) is the quadrupole coupling constant. At low temperatures, the ortho species is predicted to relax much more slowly than the para species. This has been checked for a gas of D_2 molecules [4]. At high temperatures, the two relaxation rates are expected to be almost identical. This has been checked for a gas of N_2 molecules [5].

Spin symmetry considerations should in principle be important for any molecule possessing two or more identical nuclei at equivalent sites. As another example, consider the permutation symmetry of the four protons in the methane molecule. The degenerate rotational wavefunctions characterized by a particular value of the angular momentum quantum number J may be classified according to the irreducible representations of the pure rotational group of a regular tetrahedron. For CH_4 molecules there exists a one to one correspondence between the three values of the total nuclear spin I per molecule and the three irreducible representations labelled A, T and E. In particular, the meta (A) spin symmetry species has $I = 2$, the ortho (T) species has $I = 1$ and the para (E) species has $I = 0$. As a result of centrifugal distortion the rotational degeneracy is partially removed. Nuclear spin relaxation is dominated by the spin-rotation interaction and since the spin-rotation interaction is linear in the nuclear spin operations, the only non-zero matrix elements satisfy the selection rules $\Delta I = 0$, ± 1. Therefore, in addition to a large term in $[G(t)]_{free}$ corresponding to zero frequency matrix elements between states A ↔ A and T ↔ T, there exist important terms at non-zero but low frequencies corresponding to matrix elements between states A ↔ T, T ↔ E and T ↔ T within the same J manifold. The presence of these symmetry effects are expected to manifest themselves in the T_1 behaviour. The problem of longitudinal relaxation in CH_4 gas at low densities is closely related to the problem of conversion between nuclear spin symmetry species. The conversion rate can be calculated in terms of the molecular hyperfine interactions [6]; in such a calculation it is important to include the centrifugal distortion splittings [7]. The conversion rate is several orders of magnitude faster than for a homonuclear diatomic molecule.

2. Low Density Regime

2.1 The Region of the Characteristic T_1 Minima

Studies of the longitudinal nuclear spin relaxation times in the vicinity of the characteristic minima are especially valuable as tests of the theory of nuclear magnetic relaxation in molecular gases. Once the theory is known to be correct such measurements can also provide spectroscopic information concerning the free molecule.

2.2 A Fundamental Test of Relaxation Theory

The first experiment to measure T_1 in the vicinity of its characteristic minimum at 77 K was carried out at 30 MHz [8]. This classic experiment has been repeated at other frequencies [9, 10]. At 77 K only the lowest two rotational states of the hydrogen molecule are appreciably populated. The molecules in the $J = 0$ state with associated nuclear spin $I = 0$ constitute the para species; those in the $J = 1$ state with associated nuclear spin $I = 1$ the ortho species. At this temperature 99.996 % of the ortho molecules are in the $J = 1$ state. Only the ortho molecules contribute to the nuclear magnetic resonance signal. For the H_2 molecule both the spin rotation and dipolar relaxation mechanisms are important. The reason for this is that the spin rotation contribution, which increases with the mean value of the molecular angular momentum, is particularly small whereas the dipolar contribution, which depends inversely upon the cube of the intramolecular distance, is larger than usual. The relaxation rate is

$$T_1^{-1} = (T_1^{-1})_{sr} + (T_1^{-1})_d$$

and there is no cross term.

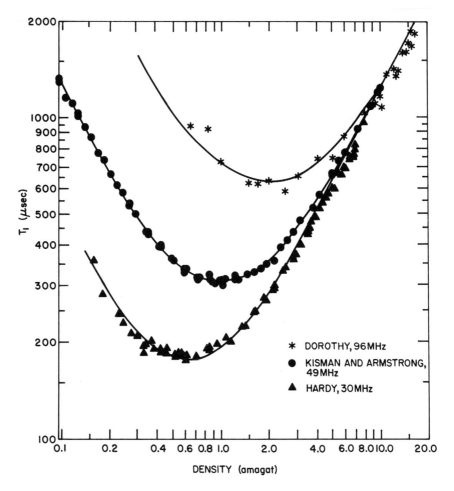

Fig. 4. T_1 data obtained in normal hydrogen at 30, 49 and 96 MHz in three separate investigations at 77.5 K

The first ab initio calculation of T_1 in hydrogen used the correlation function approach [11]. Later a quantum mechanical Boltzmann equation appropriate for the treatment of nuclear spin relaxation in dilute gases was developed [12]. Recently, the best features of both the earliers treatments have been combined to produce the most comprehensive theory yet [13, 14].

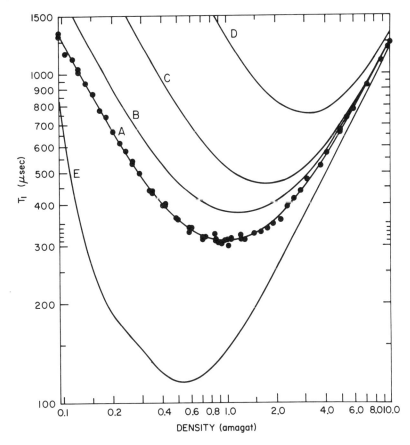

Fig. 5. Illustration of the sensitivity of T_1, in the region of the T_1 minimum for molecular hydrogen, to various aspects of the Bloom and Oppenheim (B–O) theory. [11] The curves have the following significance: (A) best fit B–O curve, (B) B–O theory with $\omega_J = 0$, (C) B–O theory neglecting the spin-rotation interaction, (D) B–O theory neglecting the dipolar interaction, (E) B–O theory with Gaussian correlation functions

For a gas of H_2 molecules at 77 K

$$T_1^{-1} = \frac{4}{3}\gamma^2 H_{sr}^2 \frac{\tau_{sr}}{1+(\omega_I-\omega_J)^2 \tau_{sr}^2} + \frac{12}{25}\gamma^2 H_d^2 \left[\frac{\tau_d}{1+(\omega_I-\omega_J)^2 \tau_d^2} + \frac{4\tau_d}{1+4(\omega_I-\omega_J)^2 \tau_d^2}\right]$$

where γ is the proton gyromagnetic ratio, $H_{sr} = 26.752$ G is the strength of the spin-rotation magnetic field and $H_d = 33.862$ G the strength of the dipolar magnetic field. The correlation times τ_{sr} and τ_d associated with the spin-rotation and dipolar fields, respectively, are related by a simple numerical factor, namely $\tau_{sr}/\tau_d = 0.6$. The difference between the nuclear Larmor frequency ω_I and the rotational Larmor frequency ω_J appears because the molecular hyperfine interaction couples states which satisfy the relation $\Delta M_I + \Delta M_J = 0$.

The data at 30, 49 and 96 MHz are shown in Fig. 4.[*] The curves are of the theoretical form and exhibit the predicted dependence on the Larmor frequency. Fig. 5 shows the sensitivity of T_1 in the region of the T_1 minimum to various aspects of the theory. Curve A represents the best fit taking both $(T_1/\rho)_{lin}$ (the value of T_1/ρ in the limit of extreme narrowing) and τ_{sr}/τ_d as adjustable parameters. The range of probable values of the parameter τ_{sr}/τ_d namely $0.59 \leqslant \tau_{sr}/\tau_d \leqslant 0{,}79$ is in satisfactory agreement with the theoretical value 0.6. Curve B shows the effect of neglecting ω_J, curve C of neglecting the spin-rotation interaction and curve D of neglecting the dipolar interaction. Finally, curve E is based on the assumption of Gaussian correlation functions. This is an important and a fundamental experiment as it provides a sensitive test to the basic ingredients of the relaxation theory.

2.3 Free Molecule Spin-Rotation Parameters

The general spin-rotation interaction Hamiltonian has the form

$$\mathcal{H}_{sr} = \hbar \vec{I} \cdot \mathbf{C}_{sr} \cdot \vec{J}$$

where $\mathbf{C}_{sr}$ is the spin-rotation tensor. For spherical top molecules of symmetry $T_d(XY_4)$ and $O_h(XY_6)$ $\mathbf{C}_{sr}$ is diagonal with components $C_\perp$ proportional to the magnetic field per unit angular momentum generated at a nuclear site by rotation about an axis perpendicular to an $X - Y$ bond axis of the molecule and $C_\parallel$ proportional to the magnetic field per unit angular momentum generated by rotations about an $X - Y$ bond axis. The relaxation rate associated with this interaction is

$$(T_1^{-1})_{sr} = 4\pi^2 \frac{2I_0 kT}{\hbar^2} \left[C_a \frac{\tau_{sr}}{1 + (\omega_I - \omega_J)^2 \tau_{sr}^2} + \frac{4}{45} C_d^2 \frac{\tau'_{sr}}{1 + (\omega_I - \omega_J)^2 \tau'^2_{sr}} \right]$$

where $C_a = \frac{1}{3}(2C_\perp + C_\parallel)$ and $C_d = C_\perp - C_\parallel$. If the two correlation times τ_{sr} and τ'_{sr} are equal

$$(T_1^{-1})_{sr} = 4\pi^2 \frac{2I_0 kT}{\hbar^2} C_{eff}^2 \frac{\tau_{sr}}{1 + (\omega_I - \omega_J)^2 \tau_{sr}^2}$$

with $C_{eff}^2 = C_a^2 + \frac{4}{45} C_d^2$.

Although molecular beam experiments provide accurate values of the average spin-rotation constant C_a for spherical top molecules such as CF_4, SiF_4 and GeF_4, such experiments do not have sufficient resolution to provide more than an upper limit for the anisotropy parameter C_d. If we believe the theoretical expression for the nuclear longitudinal relaxation rate we can use fluorine T_1 measurements through

---

[*] The amagat density is defined as the ratio of the number of molecules per unit volume at a given temperature and pressure to the number of molecules per unit volume at NTP.

the region of the characteristic minimum to obtain a value of the effective spin-rotation constant C_{eff}. Therefore, a value of C_d may be deduced from the measured values of C_a and C_{eff}.

Fig. 6 shows the density dependence of T_1 for the fluorine nuclei in CF_4, SiF_4 and GeF_4 [15]. For low density gases of these molecules, fluorine nuclear relaxation is dominated by the spin-rotation interaction. The values of C_a^2, C_{eff}^2 and C_d^2 are given in Table 1. It must be emphasized that these results are based on the assumption that the two correlation times τ_{sr} and τ'_{sr} are equal. Information from other experiments, as for example Sentfleben-Beenakker viscosity measurements in CF_4, suggest that the approximation is a reasonable one. Fig. 7 shows the variation of C_d^2 with $n = \tau'_{sr}/\tau_{sr}$ as deduced from nuclear longitudinal relaxation time measurements in GeF_4. The vertical dashed lines define the range of values of n consistent with the molecular beam results. For n near unity C_d^2 is almost independent of n and the assumption that $n = 1$ does not provide a severe limitation to the deduction of C_d^2 from these experiments.

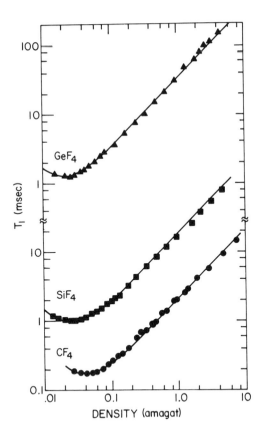

Fig. 6. T_1 data for the fluorine nuclei in the spherical top molecules GeF_4, SiF_4, CF_4 at 293 K

Table 1. Spin-rotation Parameters for Several XF$_4$ Molecules

Molecule	C_{eff}^2 (kHz)^2	C_a^2 (kHz)^2	C_d^2	
			NMR (kHz)^2	Mol. Beam (kHz)^2
CF$_4$	47.6 ± 1.1	46.5 ± 0.6	12.5 $^{+17}_{-12}$	< 200
SiF$_4$	6.30 ± 0.22	5.86 ± 0.10	4.95 ± 3.60	< 9
GeF$_4$	4.02 ± 0.10	3.53 ± 0.06	5.50 ± 1.80	< 9

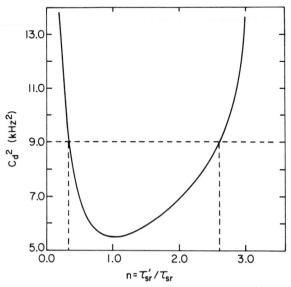

Fig. 7. A plot to illustrate the dependence of C_d^2 on the ratio $n = \tau_{sr}'/\tau_{sr}$ for GeF$_4$. The horizontal dashed line corresponds to the upper limit for C_d^2 as deduced from molecular beam studies. The vertical dashed lines define the range of values of n consistent with the molecular beam results

2.4 The Effects of Centrifugal Distortion

It has been demonstrated [2, 16] that the effects of centrifugal distortion influence the longitudinal relaxation in methane gas in the vicinity of the relaxation time minimum. Relaxation of the protons is dominated by the spin-rotation interaction. The term containing C_a (scalar term) involves only zero frequency matrix elements while the term containing C_d (tensor term) also connects states whose energies differ by a centrifugal distortion energy. A straightforward extension of the usual expression for the relaxation rate leads to the result

$$(T_1^{-1})_{sr} = 4\pi^2 \frac{2I_0 kT}{\hbar^2} C_a^2 \left[\frac{\tau_{sr}}{1 + (\omega_I - \omega_J)^2 \tau_{sr}^2} \right.$$

$$\left. + \frac{4}{45} C_d^2 \sum_K \frac{F_K \tau_{sr}^K}{1 + (\omega_I - \omega_J - \omega_K)^2 \tau_{sr}^{K2}} \right]$$

at high temperatures, if the relaxation is exponential. The F_K give a measure of the sum of squares of the matrix elements of tensor part of the interaction between states separated in energy by $\hbar\omega_K$; the F_K satisfy the normalization condition $\Sigma F_K = 1$.

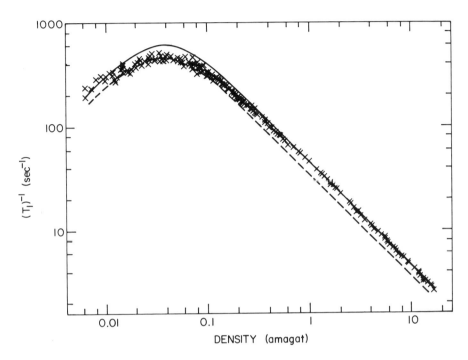

Fig. 8. T_1^{-1} data for methane at 295 K. The solid line is the theoretical prediction neglecting centrifugal distortion. The dashed line is the theoretical prediction assuming that the terms involving C_d are completely quenched

Fig. 8 is a plot of T_1^{-1} as a function of density of 30 MHz. The solid curve shows the predicted density dependence using the conventional theory (neglecting centrifugal distortion) with the known values of C_a and C_d. At the highest densities the rotational levels are broadened by an amount which is larger than the average centrifugal distortion frequency so that spin symmetry effects should not be observable. The dashed curve shows the predicted density dependence for $C_d = 0$. At the density of the T_1^{-1} maximum, the broadening of the rotational levels is much less than the average centrifugal distortion frequency and the terms involving C_d are almost completely quenched. The difference between the measured T_1^{-1} values and those predicted for $C_d = 0$ directly is related to the centrifugal distortion spectrum. The observed deviation is consistent with an average centrifugal distortion splitting in the vicinity of 200 MHz.

3. Intermediate Density Regime

3.1 The Study of Intermolecular Forces by Means of T_1 Measurements

One of the most important applications of the measurement of longitudinal nuclear spin spin relaxation times in gases is to the study of intermolecular interactions, and especially the anisotropic intermolecular potential. The density regime for which T_1 is proportional to ρ will be referred to as the intermediate density regime. It has been the most useful for these studies. Fig. 9 shows the linear variation of T_1 with density in fluorine gas at two temperatures [17]. In all theoretical approaches, the spin system is treated quantum mechanically; however, the rotational and translational motions are in some cases treated classically and in other in quantum mechanically.

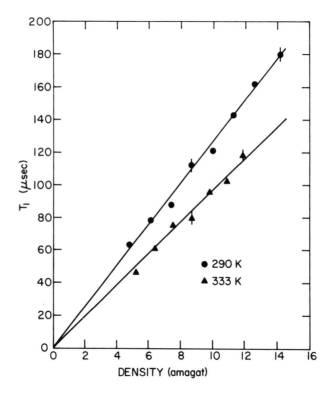

Fig. 9. T_1 data in fluroine gas taken in the intermediate density regime at two temperatures

3.2 A Simple Kinetic Theory Approach

A simple kinetic theory treatment of longitudinal nuclear spin relaxation for gases containing linear molecules has been proposed [18, 19]. The assumptions usually employed for kinetic theory treatments of dilute gases are made to calculate the appropriate correlation functions.

The relaxation rate due to the spin-rotation interaction in a dilute gas of diatomic molecules for densities for which $(\omega_I - \kappa_J)\tau_{sr} \ll 1$ is

$$(T_1^{-1})_{sr} = 4\pi^2 \frac{4}{3} \frac{I_0 kT}{\hbar^2} C_{sr}^2 \tau_{sr} .$$

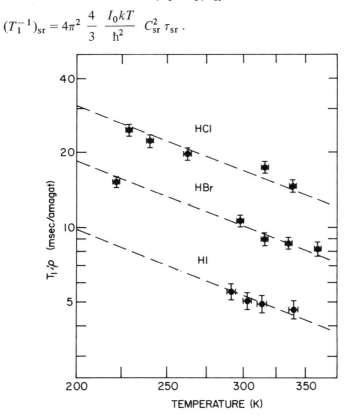

Fig. 10. A log-log plot of T_1/ρ versus T for the diatomic gases HCl, HBr and HI. The lines have slope -1.5

The relevant correlation function is the auto-correlation function $\langle \vec{J}(0) \cdot \vec{J}(t) \rangle$ of the molecular rotational angular momentum $\vec{J}(t)$. From elementary kinetic theory assuming binary collisions $\tau_{sr} = \dfrac{1}{\rho L \bar{v} \sigma_{sr}}$ where ρ is the gas density in amagat units, L is Loschmidt's number, $\bar{v}$ is the mean relative velocity of a colliding pair of molecules and σ_{sr} is an effective cross section for the transfer of angular momentum during a collision. The mean velocity is $\bar{v} = (8kT/\pi\mu)^{1/2}$ where μ is the reduced mass of a colliding pair of molecules. According to the theory, T_1 depends upon the velocity averaged cross section $\langle v\sigma_J \rangle$ which is related to σ_{sr} by $\langle v\sigma_J \rangle = \bar{v}\sigma_{sr}$. The cross section σ_J is given by

$$\sigma_J = \frac{1}{2\langle J^2 \rangle} \int_0^\infty \langle (\Delta J)^2 \rangle_i \, 2\pi b \, db$$

where $(\Delta J)^2$ is the square of the angular momentum transfer during a collision and the bracket $\langle \ \rangle_i$ implies that $(\Delta J)^2$ is averaged over the initial distribution of internal states before a collision. If the collision integral is independent of temperature, it follows from the principle of equipartition of energy that $\sigma_J \propto T^{-1}$ and therefore that T_1/ρ is proportional to $T^{-3/2}$.

Fig. 10 shows the temperature dependence of T_1/ρ for HCl, HBr and HI gas [20]. The straight lines drawn through the data have slope -1.5. It therefore seems that the collision integral at most has a very weak temperature dependence.

From the experimental (T_1/ρ) values the corresponding cross sections σ_{sr} may be deduced. For HCl and HI at 315 K the respective values are 94 and 129 Å^2. For comparison the corresponding geometric cross sections are 34 and 53 Å^2. It follows that

$$\left(\frac{\sigma_{sr}}{\sigma_{geom}}\right)_{HCl} = 2.8 \text{ and } \left(\frac{\sigma_{sr}}{\sigma_{geom}}\right)_{HI} = 2.4.$$

The large values of the cross sections σ_{sr} reflect the importance for the transfer of angular momentum of the long range dipole-dipole forces between polar molecules. The molecular dipole moment of HCl is 1.08 D and of HI is 0.42 D. In contrast, for a gas of non-polar F_2 molecules [17] σ_{sr} at 300 K is 12 Å^2, σ_{geom} is 41 Å^2 and $(\sigma_{sr}/\sigma_{geom})_{F_2} = 0.30$.

3.3 A Semiclassical Scattering Approach to the Calculation of Cross Sections

The ultimate aim of all theoretical approaches to the problem of molecular collisions in gases is to obtain information concerning the intermolecular potential. A general method has been developed for the calculation of cross sections associated with diatomic molecule-atom collisions in a semiclassical approximation [21]. That is, the relative translational motion is specified using classical mechanics, the rotational motion using quantum mechanics and the relative translational motion is assumed to be independent of the rotational motion. A potential is chosen, the scattering matrices calculated from the potential and cross sections calculated from the scattering matrices for a variety of experiments including nuclear spin relaxation. Whereas a single experiment provides some information about the potential, a combination of results from several experiments permits a better determination of the potential.

The method has been applied in detail to the HCl-Ar system [22]. The intermolecular potential $V(r, \theta)$ is a general function of the distance r between the centers of mass of the two particles and the angle θ between the direction of $\vec{r}$ and the HCl molecular axis. The potential is approximated by a truncated series of Legendre polynomials

$$V(r, \theta) = \sum_{k=0}^{k_{max}} V_k(r) P_k(\cos \theta).$$

The first term gives the isotropic potential $V_0(r)$ between HCl and Ar. This potential is represented by the modified Buckingham form

$$V_0(r) = \epsilon \frac{6\alpha}{1 - 6/\alpha} \exp\left[\alpha\left(1 - \frac{r}{r_m}\right)\right] - \frac{\epsilon}{1 - 6/\alpha}\left(\frac{r_m}{r}\right)^6$$

where the parameters for HCl-Ar are $\epsilon/k = 202$ K, $r_m = 3.805$ Å, $\alpha = 13.5$. This potential is used to calculate the classical path of the translational motion. For convenience in introducing the anisotropic terms, $V_0(r)$ is divided into attractive $V_{oa}(r)$ and repulsive $V_{or}(r)$ contributions. Since the anisotropic part of the asymptotic long range potential is closely related to the isotropic part, the attractive anisotropic potential $V_a(r, \theta)$ is expressed as a product of $V_{oa}(r)$ and a series of Legendre polynomials. Retaining only the leading terms

$$V_a(r, \theta) = V_{oa}(r) \left[1 + P1A \cdot P_1(\cos \theta) \left(\frac{r_m}{r} \right) + P2A \cdot P_2(\cos \theta) \right]$$

where $P1A$ and $P2A$ are variable parameters which include contributions from both induction and dispersion forces. Available experimental data do not justify the use of more terms in the expansion. By analogy, the repulsive anisotropic potential $V_r(r, \theta)$ is given the form

$$V_r(r, \theta) = V_{or}(r) [1 + P1R P_1(\cos \theta) + P2R P_2(\cos \theta)]$$

where $P1R$ and $P2R$ are variable parameters.

Relations between this potential and the experimental measurements of dipole absorption spectra, Raman scattering, proton magnetic relaxation, chlorine quadrupolar relaxation, rotational relaxation and sound absorption have been derived. By combining all available experimental data a single best set of parameters has been obtained: $P1A = 0.32 \pm 0.05$, $P2A = 0.24 \pm 0.10$, $P1R = 0.51 \pm 0.05$, $P2R = 0.78 \pm 0.10$.

In order to take full advantage of this theoretical advance, there is now a need for more experimental data on other diatomic molecule-atom systems.

3.4 The Special Case of Hydrogen Gas

Pure hydrogen gas provides a particularly simple system from which to obtain information about the intermolecular potential. Since at most a few rotational states are appreciably populated it is possible to carry out ab initio calculations and obtain a precise determination of the relation between T_1/ρ and the intermolecular potential in terms of a small number of parameters [11, 13, 14].

In Section 1.4 the importance of nuclear spin symmetry species was discussed. Although for hydrogen gas only the ortho molecules can be detected by nuclear magnetic resonance the measured values of T_1/ρ depend on the ortho-para concentration since an ortho-ortho collision is different from an ortho-para collision. Fig. 11 is a plot of T_1/ρ versus ortho-H_2 concentration n_0 at 77.5 K. As predicted by the theory, a linear variation is observed. The linear dependence has been checked up to room temperature [23]. From this type of data it is possible to construct the temperature dependence of $(T_1/\rho)^{oo}$ resulting solely from ortho-ortho collisions and $(T_1/\rho)^{op}$ resulting solely from ortho-para collisions. This result is illustrated in Fig. 12 along with the temperature dependence of T_1/ρ in normal hydrogen. The figure shows that it is just a coincidence that T_1/ρ in normal hydrogen is practically temperature independent from 80 to 300 K.

The leading terms in the expression for the anistropic potential between a pair of H_2 molecules are

$$V_A(r, \Omega_1, \Omega_2) = b^{(1)}(r) [P_2(\cos \theta_1) + P_2(\cos \theta_2)] + b^{(2)}(r) \sum_{q=-2}^{2} a_q Y_{2q}(\Omega_1) Y_{2q}^*(\Omega_2)$$

where $b^{(1)}(r)$ and $b^{(2)}(r)$ are functions of the separation r between the molecules and $\Omega_1 = \theta_1, \phi_1$ and $\Omega_2 = \theta_2, \phi_2$ denote the orientations of molecules 1 and 2 in a coordinate system with polar axis along $\vec{r}$. At large distances the second term is due to the quadrupole-quadrupole interaction, in which case $a_0 = 6$, $a_{\pm 1} = 1$ and $b^{(2)}(r) = 4\pi Q^2/r^5$ where Q is the quadrupole moment of the H_2 molecule. An analysis of H_2 data [24] taking the isotropic intermolecular potential as a Lennard-Jones potential, has shown that the radial dependence of $b^{(2)}(r)$ is that of the quadrupole-quadrupole interaction and that $Q = (0.57 \pm 0.03) \times 10^{-26}$ esu, in reasonable agreement with the theoretical value of 0.65×10^{-26} esu. The data were also used to distinguish between different forms of $b^{(1)}(r)$; the determination however is not unique. For example, if $b^{(1)}(r)$ is assumed to be of the form $b^{(1)}(r) = \dfrac{A}{r^n} - \dfrac{B}{r^6}$, the best fit to the data is achieved for $n = 15$ whereas if $b^{(1)}(r)$ is taken to be proportional to the isotropic potential $b^{(1)}(r) = \delta V_0(r)$, then an adequate fit is obtained for $\delta = 0.173$. In any event, it has been pointed out that the approach used neglects certain quantum mechanical effects [25] which are of particular importance for the determination of $b^{(1)}(r)$ and that the role of inelastic collisions is not adequately treated.

The most comprehensive approach to the investigation of the intermolecular potential in the hydrogen molecule — inert gas atom system involves the combination of data from a variety of experiments with numerical quantum scattering calcula-

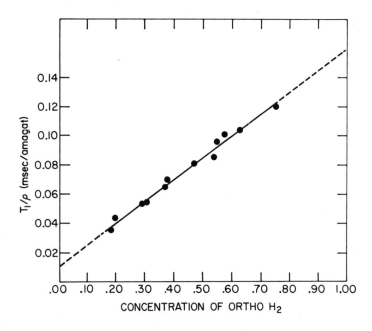

Fig. 11. Plot of T_1/ρ in hydrogen gas at 77.5 K as a function of the concentration of ortho-H_2

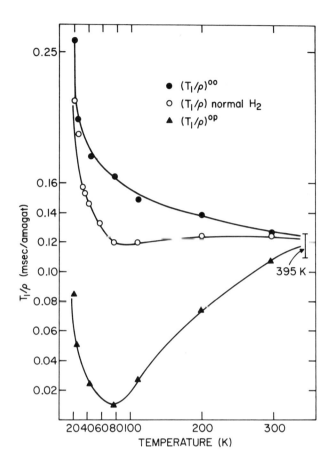

Fig. 12. Temperature dependence of T_1/ρ for hydrogen gas for three ortho-H_2 concentrations. $(T_1/\rho)^{oo}$ values result solely from ortho-ortho collisions and $(T_1/\rho)^{op}$ values from ortho-para collisions

tions [26, 27]. Note that the full quantum method is necessary only when both rotational and translational quantum numbers are small, as for example in the case of H_2–He collisions.

For H_2–He collisions the Morse-Spline fitted van der Waals potential has been used for the isotropic potential. It may be written as

$$V_0(r) = \epsilon \left\{ \exp\left[2\beta\left(1 - \frac{r}{r_m}\right)\right] - 2\exp\left[\beta\left(1 - \frac{r}{r_m}\right)\right] \right\} \quad r \leq r_1$$
$$V_0(r) = b_1 + (r - r_1)\{b_2 + (r - r_2)[b_3 + (r - r_1)b_4]\} \quad r_1 \leq r \leq r_2$$
$$V_0(r) = -\frac{C_8}{r^8} - \frac{C_6}{r^6}, \quad r \geq r_2.$$

The parameters b_i are determined by matching $V_0(r)$ and $dV_0(r)/dr$ at r_1 and r_2. In addition to longitudinal nuclear spin relaxation time data, rotational relaxation measurements, Raman line shapes and molecular beam scattering results were used to obtain the best set of potential parameters including, of course, the anisotropy parameters.

4. High Density Regime

4.1 The Passage from Intermediate to High Densities

In high density gases for spin 1/2 nuclei it is generally necessary to use experimental means to deduce the relative importance of the various relaxation mechanisms — intramolecular spin-rotation, intramolecular dipolar, intermolecular dipolar. Many papers have been written about this problem in connection with liquids. The most comprehensive work of this type concerning the passage from the intermediate density gas to the high density gas and the liquid is that of Trappeniers and his co-workers [28, 29]. Their studies involve proton T_1 measurements in methane and its deuterated modifications. Usually, however, for nuclei with spin $I > 1/2$, at sites where the electric field gradient is non zero, the intramolecular quadrupolar relaxation mechanism dominates.

Table 2. Intensity Distribution within the ν_3 Vibration-Rotation Spectrum of a CH_4-He Mixture

ρ (Amagat units)	$\dfrac{I^Q_{exp}}{I^Q_{calc}}$
430	1.00
876	1.10
955	1.26
1095	1.43
1218	1.50

At high densities account must be taken of two distinct types of effects: (1) the effect of three-body and higher order collisions (2) the effect of overlapping rotational levels and the collapse of rotational spectra. The experimental data presently available at densities for which three-body collisions are important are not adequate to provide a critical test of the existing theories taking into account the effects of a realistic potential. Evidence for the collapse of rotational spectra in gases dense enough that the molecular collision frequency is of the order of the rotational frequency has been observed in a variety of experiments. Fig. 13 shows the infrared ν_3 vibration-rotation band of methane in a CH_4–He mixture at a density of 1218 amagat units [30]. The contour is separated into "normal" P and R branches ($\Delta J = \pm 1$) and an anomalous Q branch ($\Delta J = 0$). Table 2 illustrates the onset of the collapse of the spectrum. The physical origin of the collapse may be understood in the following manner. The P- and R-branch lines can be divided into pairs. Each pair can be associated with a specific classical rotation frequency; the two lines correspond to rotation in opposite senses. Following a collision which reorients the angular momentum, the resultant rotation may be described as a superposition of the original state and the state which corresponds to rotation in the opposite sense. That is, the collision transfers some intensity between a P and an R branch line. Once the collision rate becomes comparable to the rotation frequency, each pair of rotating and counter-rotating components tends to produce a component at zero frequency. That is, the P and R branches collapse as their intensity is transferred to the Q branch. This type of motional narrowing will undoubtedly influence the behaviour of T_1.

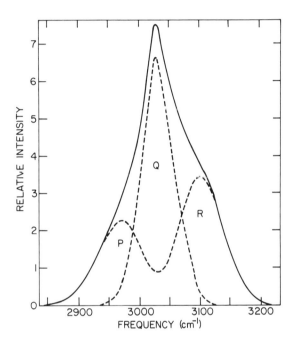

Fig. 13. The infrared ν_3 rotation-vibration band of methane in a CH_4-He mixture at a density of 1218 amagat units, showing a calculated separation of the intensity distribution into "normal" P and R branches ($\Delta J = \pm 1$) and an anomalous Q branch ($\Delta J = 0$)

4.2 A Test of the Enskog Theory

Enskog has developed a kinetic theory of dense gases based on a hard sphere model. In the model two effects are considered: (1) the modification of the collision frequency from its dilute gas value, (2) the collisional transfer of momentum and energy through the intermolecular forces. This theory is easily adapted to allow a prediction of the density dependence of longitudinal nuclear spin relaxation times. For this problem only the collision frequency effect is important because the nuclear relaxation arises almost entirely from intramolecular mechanisms. The relaxation process is essentially free from all collisional transfer effects which complicate the interpretation of transport coefficient measurements. The Enskog correction factor $\chi(\rho)$ due to the excluded volume of hard spheres is

$$\chi(\rho) = \frac{3}{2\pi\rho\sigma^3} \left[\frac{P}{\rho kT} - 1 \right]$$

where σ is the hard sphere diameter and P the gas pressure.

It is known from Raman scattering measurements that for hydrogen there is negligible overlap of the Q branch by the O and S branches ($\Delta J = \pm 2$) even in the highly compressed gas and liquid. Therefore, the interpretation of T_1 measurements in dense hydrogen gas is not complicated by the effect of overlapping rotational levels. A study of the density dependence of T_1 provides a unique opportunity to test the validity of the Enskog correction to the collision frequency in a dense gas [*31*].

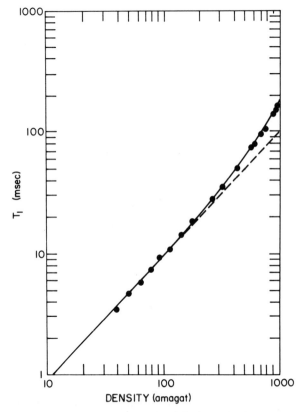

Fig. 14. A comparison of the density dependence of T_1 data in hydrogen gas at 195 K with the Enskog theory prediction of the density dependence of the collisional frequency (solid curve). The dashed line is the extrapolation of the collisional frequency in the intermediate density regime

Fig. 14 shows the density dependence of T_1 for normal H_2 at 195 K. At the highest density the gas is $\sim 25\%$ more dense than the normal liquid. The deviation from linearity at high densities is obvious. The solid line is a theoretical curve based on the Enskog correction to the collision frequency for a hard sphere diameter of 2.5 ± 0.1 Å. The dashed line is the extrapolated dilute gas behaviour. At the highest density studies the two curves differ by a factor of two. The deduced hard sphere diameter agrees well with the value 2.4 ± 0.1 Å obtained when the hard sphere equation of state is used to fit the equation of state data at 600 K.

The Enskog theory also accounts for the data at 298 K. However, at 77 K slight deviations begin to appear and, at the highest density studies in compressed liquid hydrogen at 23.5 K [*32*], the Enskog corrected collision frequency accounts for only about 2/3 of the difference between the uncorrected collision frequency prediction and the experimental values.

The most striking result of this analysis is not that it fails at low temperatures but rather that it works so well over such a large range of temperature and density.

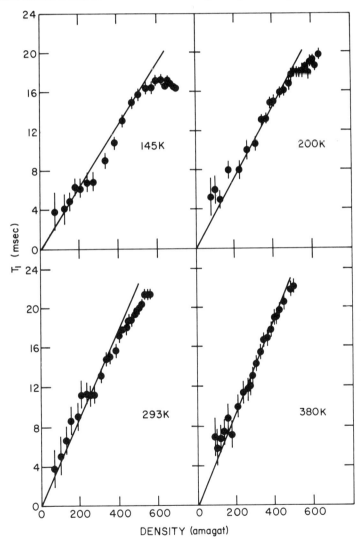

Fig. 15. T_1 data in nitrogen gas at four temperatures. Note the deviation from the linear trend at high densities

4.3 The Effect of Overlapping Rotational Levels

Fig. 15 shows the density dependence of T_1 in nitrogen gas at 145, 200, 293 and 380 K [5]. In nitrogen gas the O and S branches of the Raman spectrum overlap the Q branch even at modest densities. Note that the deviation from linearity at high densities is in the opposite sense to that predicted by the Enskog theory and observed in hydrogen gas. An explanation of the data for N_2 can be given in terms of the effect of overlapping rotational levels. At high temperatures in the short correlation time limit $(\omega_I - \omega_J)\tau_q \ll 1$

$$(T_1^{-1})_q = \frac{3}{32}\left(\frac{eqQ}{\hbar}\right)^2 \tau_q[1 + 3p\exp(p)E(p)]$$

where

$$E(p) = \int_p^\infty \frac{\exp(-x)}{x}\,dx \text{ and } \frac{1}{p} = \frac{16BkT}{\hbar}\quad \tau_q^2 = 16\langle J^2\rangle(B\tau_q)^2$$

with B the rotational constant of the nitrogen molecule and p a dimensionless parameter. At intermediate densities, $B\tau_q \ll 1$ and there is no overlapping of rotational levels. In this limit ρ approaches infinity, p exp(p) becomes negligible and

$$(T_1^{-1})_q = \frac{3}{32}\left(\frac{eqQ}{\hbar}\right)^2 \tau_q\,.$$

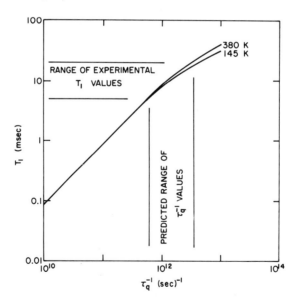

Fig. 16. A plot of the theoretical relation between T_1 and τ_q^{-1}. The range of experimentally measured T_1 values and the corresponding τ_q^{-1} values are indicated.

At high densities p decreases and becomes unity when the mean squared splitting of rotational states satisfying $\Delta J = \pm 2$ is equal to the width of the rotational states due to collisions. If all of the rotational lines overlapped with each other, P would approach zero and $pE(p)$ would approach unity. In this limit

$$(T_1^{-1})_q = \frac{3}{32}\left(\frac{eqQ}{\hbar}\right)^2 4\tau_q$$

which is a factor four larger than at intermediate densities. That is, quenching no longer occurs.

Fig. 16 is a plot of $(T_1)_q$ versus τ_q^{-1} based on the relaxation equation given above. The range of experimental T_1 values falls in the range of τ_q^{-1} values for which the departure from linearity is appreciable, corresponding to the region of non-negligible overlap of the rotational levels.

To date there are no conclusive nuclear magnetic resonance studies that illustrate the effect on the longitudinal nuclear spin relaxation time of the collapse of rotational infrared spectra.

References

[1] Gordon, R. G.: *Advances in Magnetic Resonance*, Vol. 3, p. 1. New York: Academic Press 1968.
[2] Bloom, M.: *MTP International Review of Science, Magnetic Resonance*, p. 1. London: Butterworths 1972.
[3] Abragam, A.: *The Principles of Nuclear Magnetism*. Oxford: 1961.
[4] Hardy, W. N.: Ph. D. Thesis, University of British Columbia, 1965.
[5] Speight, P. A., Armstrong, R. L.: Can. J. Phys. 47, 1475 (1969).
[6] Curl, R. F., Jr., Kaspar, J. V. V., Pitzer, K. S.: J. Chem. Phys. 46, 3220 (1967).
[7] Ozier, I., Yi, P.: J. Chem. Phys. 47, 5458 (1967).
[8] Hardy, W. N.: Can J. Phys. 44, 265 (1966).
[9] Dorothy, R. G.: Ph. D. Thesis, University of British Columbia, (1967).
[10] Kisman, K. E., Armstrong, R. L.: Can J. Phys. 52, 1555 (1974).
[11] Bloom, M., Oppenheim, I.: *Intermolecular Forces*, p. 549. New York: Interscience Publ. 1967.
[12] Chen, F. M., Snider, R. F.: J. Chem. Phys. 48, 3185 (1968).
[13] McCourt, F. R., Raidy, T. E., Festa, R., Levi, A. C.: Can J. Phys. (to be published) (1975).
[14] McCourt, F. R., Raidy, T. E., Rudensky, T., Levi, A. C.: Can. J. Phys. (to be published) (1975).
[15] Courtney, J. A., Armstrong, R. L.: Can J. Phys. 50, 1252 (1972).
[16] Beckmann, P. A., Thesis, Ph. D.: University of British Columbia (1975).
[17] Courtney, J. A., Armstrong, R. L.: J. Chem. Phys. 52, 2158 (1970).
[18] Gordon, R. G.: J. Chem. Phys. 42, 3658 (1965).
[19] Gordon, R. G.: J. Chem. Phys. 44, 228 (1966).
[20] Tward, E., Armstrong, R. L.: J. Chem. Phys. 47, 4068 (1967).
[21] Neilsen, W. B., Gordon, R. G.: J. Chem. Phys. 58, 4131 (1973).
[22] Neilsen, W. B., Gordon, R. G.: J. Chem. Phys. 58, 4149 (1973).
[23] Lipsicas, M., Hartland, A.: Phys. Rev. 131, 1187 (1963).
[24] Lalita, K., Bloom, M., Noble, J. D.: Can J. Phys. 47, 1355 (1969).
[25] Riehl, J. W., Kinsey, J. L., Waugh, J. S.: J. Chem. Phys. 46, 4546 (1967).
[26] Riehl, J. W., Fisher, C. J., Baloga, J. D., Kinsey, J. L.: J. Chem. Phys. 58, 4571 (1973).
[27] Shafer, R., Gordon, R. G.: J. Chem. Phys. 58, 5422 (1973).
[28] Gerritsma, C. J., Oosting, P. H., Trappeniers, N. J.: Physica 51, 381 (1971).
[29] Oosting, P. H., Trappeniers, N. J.: Physics 51, 395 (1971).
[30] Armstrong, R. L., Thesis, Ph. D.: University of Toronto (1961).
[31] Gordon, R. G., Armstrong, R. L., Tward. E.: J. Chem. Phys. 48, 2655 (1968).
[32] Lipsicas, M., Hartland, A.: J. Chem. Phys. 44, 2839 (1966).

Spin-Lattice Relaxation in Nematic Liquid Crystals Via the Modulation of the Intramolecular Dipolar Interactions by Order Fluctuations*)

R. Blinc

Contents

I. Introduction 97
II. Nematic Fluctuations in the Isotropic Phase $(T > T_N)$ 97
III. Nematic Phase $(T < T_N)$ 102
References 111

I. Introduction

It is well known that nuclear spin-lattice relaxation time (T_1) studies may provide valuable information on the dynamics of liquid crystals [1–6]. The onset of long range orientational order, on going from the isotropic liquid (I) into the nematic (N) phase and the occurrence of one or more dimensional translational ordering, on going from the nematic into the various smectic (Sm) liquid crystalline phases, may have profound effects on T_1. In the general case we may expect several competing relaxation mechanisms.

$$\left(\frac{1}{T_1}\right)_{\text{exp}} = \left(\frac{1}{T_1}\right)_{\text{nematic order}} + \left(\frac{1}{T_1}\right)_{\text{smectic order}} + \left(\frac{1}{T_1}\right)_{\text{self-diffusion}} + \ldots \quad (\text{I.1})$$

which may be separated due to their different angular, frequency and temperature dependences [5–9].

Here we wish to analyze the relaxation mechanism arising from the modulation of the intramolecular dipolar interactions by nematic order waves and try to determine how the corresponding T_1 depends on frequency, temperature and orientation.

II. Nematic Fluctuations in the Isotropic Phase $(T > T_N)$

Let us first analyze the contribution of nematic order fluctuations to the nuclear spin-lattice relaxation rate in the isotropic liquid phase above the isotropic-nematic transition temperature (T_N).

Following de Gennes [1], the molecules are still locally parallel to each other in the isotropic phase not too far from T_N. The free energy density $g(S, T)$ of a given non-

---
*) Presented at the NMR Summer School, Waterloo, June 22–28, 1975.

equilibrium state above T_N can be expanded [1] in powers of the nematic order parameter $S = S(\vec{r}, t)$:

$$g(S, T) = g_0 + \frac{1}{2} A S^2 + \frac{1}{2} C (\nabla S)^2 + \ldots \tag{II.1}$$

where $S = \frac{1}{2} \langle (3 \cos^2\theta - 1) \rangle$ and θ stands for the angle between the longest molecular axis and the direction of local nematic order $\vec{N}(\vec{r}, t)$. Here cubic and higher order terms are neglected.

In the above expansion C is assumed to be a positive elastic constant ($C > 0$) and the coefficient A should vary with temperature as:

$$A = a(T - T_N)^\nu, \quad a > 0. \tag{II.2}$$

As long as $A > 0$, i.e., $T > T_N$, the equilibrium value of S is zero.

Introducing the Fourier components of the nematic order parameter fluctuations $S(\vec{q}, t)$ as new variables

$$S(\vec{r}) = (V^{-1/2}) \cdot \mathrm{Re} \sum_{\vec{q}} S(\vec{q}) \cdot e^{i\vec{q} \cdot \vec{r}} \tag{II.3}$$

the free energy

$$F = \int g \, dV \tag{II.4}$$

becomes in this approximation a simple quadratic function of $S(\vec{q}, t)$:

$$F - F_0 = \frac{1}{2} \sum_{\vec{q}} A |S(\vec{q})|^2 [1 + q^2 \xi^2]. \tag{II.5}$$

The coherence length ξ

$$\xi^2 = C/A \tag{II.6}$$

measures the distance over which the local nematic order persists in the isotropic phase and diverges as $T \to T_N$.

The characteristic time $\tau_{\vec{q}}$ for a nematic order fluctuation $S(\vec{q}, t)$ of wave vector $\vec{q}$ to relax back to equilibrium can be obtained from the Landau-Khalatnikov equation:

$$\frac{d|S(\vec{q})|}{dt} = -\Gamma \cdot \frac{dF}{d|S(\vec{q}, t)|} = -\Gamma \cdot A(1 + q^2 \xi^2) \cdot |S(\vec{q})| \tag{II.7}$$

as

$$\tau_{\vec{q}} = \frac{\Gamma^{-1}}{A(1 + q^2 \xi^2)}. \tag{II.8}$$

Here Γ is a kinetic coefficient which does not vary much with temperature and Γ^{-1} represents an average viscosity of the medium. It should be noted that for long wavelength fluctuations $\tau_{q=0} \to \infty$ as $T \to T_N$.

One should point out that the above discussion has been somewhat oversimplified as the nematic order parameter is not a scalar quantity but a symmetric second rank tensor of zero trace with five independent components. In principle there are thus five independent order parameter relaxation modes [4] and not a single one. However, since above T_N all these modes are degenerate, the above treatment is still adequate.

Let us now consider the effect of these order fluctuations on a nuclear probe consisting — like the two ortho protons on a benzene ring — of two like spin 1/2 nuclei with a fixed separation distance $\vec{r}$ and a gyromagnetic ratio γ.

The spin Hamiltonian of such a probe in an external magnetic field $\vec{H}_0 \| z$ is the sum of a Zeeman and a dipolar term [7]

$$\mathcal{H} = \mathcal{H}_Z + \mathcal{H}_D \tag{II.9}$$

where

$$\mathcal{H}_D = \mathcal{H}_{12} = -\frac{(\hbar\gamma)^2}{r^3} [3(\vec{I}_1 \cdot \vec{n})(\vec{I}_2 \cdot \vec{n}) - \vec{I}_1 \vec{I}_2] \tag{II.10}$$

and $\vec{n}$ is a unit vector parallel to $\vec{r}$ which is assumed to be parallel to the longest molecular axis.

In the isotropic liquid the direction of the longest molecular axis (and hence of $\vec{n}$) rapidly changes with time, making $\mathcal{H}_D$ time dependent. There are two different types of molecular motion which are important in our case: (i) rotation of a single molecule, which occurs on a time scale of 10^{-11} sec, and (ii) change in the magnitude and orientation of the local nematic order $S(\vec{r}, t)$ — or, which is the same, rotation and deformation of a nematic short range order cluster of size ξ — which occurs on a time scale which is slower by several orders of magnitude than the rotation of a single molecule.

Both motions are fast compared to the energy level splittings of $\mathcal{H}_D$ and the nucleus sees only an averaged dipolar Hamiltonian.

Let us take first into account the fast single particle motion. Averaging over all orientations of $\vec{n}$ due to molecular reorientations inside a nematic short range order cluster we obtain:

$$\overline{n_z^2} = \frac{1}{3} + \frac{2}{3} S \tag{II.11a}$$

$$\overline{n_x^2} = \overline{n_y^2} = \frac{1}{3} - \frac{1}{3} S \tag{II.11b}$$

and

$$\overline{n_x n_y} = \overline{n_x n_z} = \overline{n_y n_z} = 0 \tag{II.11c}$$

where $S = S(\vec{r}, t)$ is the amount of local nematic order inside a cluster. Here we have assumed that the nematic short range ordering occurs along the z axis and that there is no preferential ordering in the x–y plane. This seems to be supported by the magnetic birefringence data [1]. For a time which is long compared to the single molecule rotation time but short compared to the life time of the nematic short range ordering, the individual molecules are thus oriented at an angle θ with respect to $\vec{H}_0$.

Hence we find

$$\overline{\mathcal{H}_D(t)} = \hbar\gamma H_L S(t) [\vec{I}_1 \vec{I}_2 - 3\vec{I}_{z1} \vec{I}_{z2}] \tag{II.11d}$$

where $H_L = \gamma\hbar/r^3$.

The average over the short range nematic order fluctuations yields of course $\langle \mathcal{H}_D(t) \rangle = 0$ for $T > T_N$, as on a time scale which is long compared to the life time of a nematic cluster $\langle \cos^2\theta \rangle = 1/3$ and $\langle S(t) \rangle = 0$. The spectral density of the time fluctuations of the short range nematic order $\langle S(0) S(t) \rangle$ and the resulting modulation of the nuclear dipole-dipole interactions gives rise however, to a non-vanishing contribution to Zeeman spin-lattice relaxation rate T_1^{-1} as well as to the linewidth T_2^{-1} and the rotating frame spin-lattice relaxation rate $T_{1\rho}^{-1}$. We find

$$J(\omega) = \int_{-\infty}^{+\infty} \overline{(\langle m|\mathcal{H}_D(0)|k\rangle \langle k|\mathcal{H}_D(t)|m\rangle)}\, e^{-i\omega t} dt =$$
$$= \frac{1}{4} (\gamma H_L)^2 \int_{-\infty}^{+\infty} \langle S(0)S(t)\rangle\, e^{-i\omega t} dt . \tag{II.12}$$

This expression gives as well the nematic short range order fluctuation contribution [7] to the quadrupole spin-lattice or spin-spin relaxation rate of a spin 1 nucleus – like ^{14}N or ^{2}H – if $(\gamma H_{Loc})^2$ is replaced by $(e^2qQ/2)^2$ where q is the largest principal value of the electric field gradient tensor at the nuclear site and Q is the nuclear quadrupole moment.

Using expressions (II.5), (II.7) and (II.8), the autocorrelation function of S can be expressed as:

$$\langle S(0) S(t) \rangle = \frac{1}{V} \sum_{\vec{q}} \langle |S(\vec{q})|^2 \rangle\, e^{-t/\tau_{\vec{q}}} . \tag{II.13}$$

The expression for $J(\omega)$ thus becomes:

$$J(\omega) = \frac{1}{4} (\gamma H_L)^2 \frac{1}{V} \sum_{\vec{q}} J_{\vec{q}}(\omega) \tag{II.14}$$

where the spectral density $J_{\vec{q}}(\omega)$, of order fluctuations with wave vector $\vec{q}$ is given by

$$J_{\vec{q}}(\omega) = \langle |S(\vec{q})|^2 \rangle \frac{2\tau_{\vec{q}}}{1 + (\omega\tau_{\vec{q}})^2} . \tag{II.15}$$

Using the equipartition theorem and expression (II.5), then mean square fluctuation in the nematic order parameter with wave vector $\vec{q}$ is obtained as:

$$\langle |S(\vec{q})|^2 \rangle = \frac{kT}{A(T)(1+q^2\xi^2)} \tag{II.16}$$

so that $J_{\vec{q}}(\omega)$ becomes:

$$J_{\vec{q}}(\omega) = \frac{kT}{\Gamma^{-1}} \cdot \frac{2}{\omega^2 + [C\Gamma(q^2 + 1/\xi^2)]^2} \tag{II.17}$$

Replacing the summation over $\vec{q}$ by an integration and using for the density of $\vec{q}$ states $\frac{V}{(2\pi)^3}$ we find [7]:

$$J(\omega) = \frac{(\gamma H_L)^2}{2 \cdot (2\pi)^3} \cdot \frac{kT}{\Gamma^{-1}} \int_0^{q_m} \frac{4\pi q^2 dq}{\omega^2 + [C\Gamma(q^2 + 1/\xi^2)]^2} \tag{II.18}$$

The integral

$$I = \int_0^{q_m} \frac{4\pi q^2 dq}{\omega^2 + [C\Gamma(q^2 + 1/\xi^2)]^2} = \frac{2\pi}{(C\Gamma)^{3/2}} \int_0^{x_m} \frac{\sqrt{x}\, dx}{\omega^2 + (x + r)^2} \tag{II.19}$$

with $r = C\Gamma/\xi^2 = \omega_C$ can be solved in a closed form. The result is however rather clumsy. It depends on three parameters: the nuclear Larmor frequency ω, the temperature dependent "correlation frequency" $r = C\Gamma/\xi^2$ and the "cut-off" frequency $x_m = C \cdot \Gamma \cdot q_m^2$, where $q_m \approx \pi/\ell$ and ℓ is of the order of a molecular length (20–30 Å). As the coherence length ξ was found [7] to be of the order of 200–400 Å a few degrees above T_N, $r \ll x_m$ in the region of interest. The nuclear Larmor frequencies in a magnetic field of the order of $H_0 = 10^4$ Gauss are also much smaller than x_m. Thus we can make the approximation $x_m \to \infty$, and the result simplifies to:

$$(T_1^{-1}) \propto I = \frac{2\pi}{(C\Gamma)^{3/2}} \cdot \frac{\sin\{\frac{1}{2} \tan^{-1}(\omega/r)\}}{\sin\{\tan^{-1}(\omega/r)\}} \cdot \frac{1}{(\omega^2 + r^2)^{1/4}} \tag{II.20}$$

showing the dependence of T_1^{-1} on the nuclear Larmor frequency ω and the correlation length $\xi = \sqrt{C\Gamma/r}$.

The above expression becomes particularly simple in two limiting cases:

(i) $\omega \ll r = \omega_C$,

(ii) $\omega \gg r = \omega_C$.

In the first case of "small" nuclear Larmor frequencies we find:

$$I \propto \frac{1}{\sqrt{r}} \propto \xi \tag{II.21a}$$

so that T_1 is strongly temperature dependent [7]:

$$T_1 \propto \xi^{-1} \propto (T - T_N)^{\nu/2} \neq f(\omega). \tag{II.21b}$$

In the second case of "large" nuclear Larmor frequencies

$$I \propto \frac{1}{\sqrt{\omega}} \tag{II.21c}$$

and T_1 is frequency but not strongly temperature dependent:

$$T_1 \propto \sqrt{\omega} \neq f(T - T_N). \tag{II.21d}$$

This second case always occurs in the immediate vicinity of T_N, where ξ becomes large and r small.

The result (II.21b) has been already derived and experimentally verified in PAA by Cabane and Clark [7]. The frequency dependence of T_1 in the isotropic phase up to 15 °K above T_N — as given by expression (II.21d) — has been as well already observed [3].

It should be noted that the correlation length ξ can be obtained from Eq. (II.20) not only by measuring the temperature dependence of T_1^{-1} at fixed frequency but also by measuring the frequency dependence of (T_1^{-1}) at fixed temperature. Both methods work, however, only as long as the self-diffusion contribution — see Eq. (1) — does not dominate T_1^{-1}.

III. Nematic Phase ($T < T_N$)

The fluctuations in the magnitude of the nematic ordering occur for $T < T_N$ at much higher frequencies than the fluctuations in the orientation of the local nematic order [1, 4]. The fluctuations most important for nuclear relaxation are those having correlation times of the order of the Larmor period, ω^{-1}. This limits the modes of interest to coherent angular reorientation modes with wavelengths of the order of $10^2 - 10^3$ Å.

In view of the long wavelengths involved the elastic continuum theory [1] of liquid crystals can be used. In this theory the distorted state of a nematic crystal is described in terms of a vector field $\vec{N}(\vec{r}, t)$. The orientation of the unit vector $\vec{N}$ varies slowly with $\vec{r}$.

In nematic systems the fluctuations in the nematic director $\vec{N}(\vec{r}, t)$ are produced [1] by "splay" (K_1), "twist" (K_2) and "bend" (K_3) type elastic deformations.

The corresponding change in the free energy density due to these deformations is [1]

$$g_{d,n} = \frac{1}{2} K_1 (\operatorname{div} \vec{N})^2 + \frac{1}{2} K_2 (\vec{N} \cdot \operatorname{curl} \vec{N})^2 + \frac{1}{2} K_3 (\vec{N} \times \operatorname{curl} \vec{N})^2. \tag{III.1}$$

Let us now assume that the nematic director thermally fluctuates around an average direction $\vec{N}_0$ which is parallel to the z axis:

$$\vec{N}(\vec{r}, t) = \vec{N}_0 + \delta \vec{N}(\vec{r}, t) \tag{III.2}.$$

$\delta \vec{N} = (\delta N_x, \delta N_y, 0)$ is normal to $\vec{N}_0$ as $(\vec{N})^2 = 1$. The free energy of deformation now becomes [1]

$$F_{d,n} = \int g_{d,n} dV = \frac{1}{2} \int dV \left\{ K_1 \left(\frac{\partial N_x}{\partial x} + \frac{\partial N_y}{\partial y} \right)^2 + K_2 \left(\frac{\partial N_x}{\partial y} - \frac{\partial N_y}{\partial x} \right)^2 + K_3 \left[\left(\frac{\partial N_x}{\partial z} \right)^2 + \left(\frac{\partial N_y}{\partial z} \right)^2 \right] \right\}. \tag{III.3}$$

Introducing the Fourier components

$$\delta \vec{N}_x(\vec{r}) = \frac{1}{\sqrt{V}} \sum_{\vec{q}} N_x(\vec{q}) e^{-i\vec{q}\vec{r}} \tag{III.4a}$$

$$\delta \vec{N}_y(\vec{r}) = \frac{1}{\sqrt{V}} \sum_{\vec{q}} N_y(\vec{q}) e^{-i\vec{q}\vec{r}} \tag{III.4b}$$

the free energy (III.3) becomes

$$F_{d,n} = \frac{1}{2} \sum_{\vec{q}} \{K_1 |N_x(\vec{q})q_x + N_y(\vec{q})q_y|^2 + K_2 |N_x(\vec{q})q_y - N_y(\vec{q})q_x|^2$$

$$+ K_3 q_z^2 [|N_x(\vec{q})|^2 + |N_y(\vec{q})|^2]\} \tag{III.5a}$$

or

$$F_{d,n} = \frac{1}{2} \sum_{\vec{q}} \sum_{\alpha=1}^{2} |N_\alpha(\vec{q})|^2 (K_3 q_z^2 + K_\alpha q_\perp^2) \tag{III.5b}$$

where $q_\perp^2 = q_x^2 + q_y^2$ and where a linear transformation was made $-[N_x(\vec{q}), N_y(\vec{q})] \to [N_1(\vec{q}), N_2(\vec{q})]$ to diagonalize the quadratic form (III.5a). Here $N_1(\vec{q})$ is the component of $N(\vec{q})$ in the direction of the projection of $\vec{q}$ on the x-y plane, and $N_2(\vec{q})$ as well lies in the x-y plane but is perpendicular to $N_1(\vec{q})$.

The mean square fluctuation in the orientation of the director is now obtained from (III.5b) with the help of the equipartition theorem as:

$$\langle |N_\alpha(\vec{q})|^2 \rangle = \frac{kT}{(K_3 q_z^2 + K_\alpha q_\perp^2)}, \quad \alpha = 1,2 \tag{III.6}$$

and the relaxation times $\tau_\alpha(\vec{q})$ for the two normal modes describing the regression of $N_1(\vec{q})$ and $N_2(\vec{q})$ back to equilibrium are obtained from the Landau-Khalatnikov equation (II.7) as:

$$\tau_\alpha^{-1}(\vec{q}) = \frac{K_\alpha q_\perp^2 + K_3 q_z^2}{\eta_\alpha}, \quad \alpha = 1,2 \tag{III.7}$$

where $\eta_\alpha = \Gamma_\alpha^{-1}$ is an appropriate viscosity. It should be noted that in contrast to the result for the order fluctuation relaxation rate τ_q^{-1} (II.8) in the isotropic phase, the frequencies $\omega_\alpha = i \tau_{q,\alpha}^{-1}$ of the two angular reorientation modes go to zero in the long wave length limit ($q \to 0$) in the nematic phase. The two modes responsible for spin-lattice relaxation via order fluctuations in the nematic phase are thus the Goldstone modes of the isotropic-nematic transition [4].

The Zeeman spin-lattice relaxation rate T_1^{-1} of our nuclear probe consists as before of two spin 1/2 nuclei with a fixed separation distance r. The dipole-dipole interactions of the spin pairs are modulated by thermal fluctuations in the orientation of the internuclear vector $\vec{r}$ with respect to the external magnetic field $\vec{H}_0$. The relaxation rate is

$$T_1^{-1} = \frac{9}{8} \gamma^4 \hbar^2 r^{-6} [J_1(\omega_0) + J_2(2\omega_0)] \tag{III.8}$$

where $\omega_0 = \gamma H_0$ is the nuclear Larmor frequency and $J_h(p\omega)$ the spectral density of the autocorrelation function of the time dependent angular part $F_h(t)$ of the dipole-dipole interaction Hamiltonian:

$$J_h(p\omega) = \int_{-\infty}^{+\infty} \langle F_h(0) F_h(t) \rangle e^{-ip\omega t} dt. \tag{III.9}$$

Here

$$F_0(t) = 1 - 3\cos^2 \vartheta, \tag{III.10a}$$
$$F_1(t) = \cos \vartheta \cdot \sin \vartheta \cdot e^{i\varphi}, \tag{III.10b}$$
$$F_2(t) = \sin^2 \vartheta \cdot e^{2i\varphi} \tag{III.10c}$$

with ϑ and φ representing the polar angles specifying the direction of the internuclear vector $\vec{r}$ in the laboratory frame where the z-axis coincides with the direction of the applied magnetic field $\vec{H}_0$. For sake of simplicity we shall further assume that the internuclear vector is parallel to the long molecular axis, as this very nearly the case for the two orthoprotons in the benzene rings.

The rotating frame spin-lattice relaxation rate, on the other hand, is given by

$$T_{1\rho}^{-1} = \frac{9}{8} \gamma^4 \hbar^2 r^{-6} \left[\frac{1}{4} J_0(2\omega_1) + \frac{5}{2} J_1(\omega_0) + \frac{1}{4} J_2(2\omega_0) \right] \tag{III.11}$$

where $\omega_1 = \gamma H_1$ and H_1 is the amplitude for the radiofrequency field.

Since we are interested in the angular dependence of T_1 let us evaluate the correlation functions $\langle F_h(0)F_h(t)\rangle$ in a system $(z'\|\vec{N}_0)$ where $\vec{N}_0$ and $\vec{H}_0\|z$ do not coincide. The relation between $F_h(t)$ in the magnetic field fixed frame (Σ) and the primed $\vec{N}_0$ fixed frame (F_h') has been given by Doane, Tarr and Nickerson [4] and can be expressed in terms of the Euler angles $(\Phi', \Theta' = \Delta, \Psi')$.

As the instantaneous preferred direction of orientation of the molecular axes $\vec{N}(\vec{r}, t)$ thermally fluctuates around $\vec{N}_0$, a second infinitesimal transformation of the functions $F_h'(t)$ from $\vec{N}_0$ to the doubly primed frame — $F_h''(t)$ — of $\vec{N}(\vec{r}, t)$ has to be made. Averaging over the single particle molecular motion in the doubly primed frame we find [4]

$$\langle F_0''(0) F_0''(t) \rangle = 4S^2, \tag{III.12a}$$
$$\langle F_1''(0) F_1''(t) \rangle = 0, \tag{III.12b}$$
$$\langle F_2''(0) F_2''(t) \rangle = 0, \tag{III.12c}$$

so that

$$\langle F_h(0)F_h^*(t)\rangle = f_h(\Delta) \langle F_1'(0)F_1'^*(t)\rangle = f_h(\Delta)S^2 \cdot \langle \delta\vec{N}(\vec{r},0)\delta\vec{N}(\vec{r},t)\rangle \tag{III.13}$$

where $S = \langle \frac{3}{2} \cos^2\theta - \frac{1}{2} \rangle$ is the nematic order parameter and Δ the angle between $\vec{N}_0$ and $\vec{H}_0$. The functions $f_h(\Delta)$ are given by

$$f_0(\Delta) = 18 (\cos^2\Delta - \cos^4\Delta), \tag{III.14a}$$
$$f_1(\Delta) = \frac{1}{2} (1 - 3\cos^2\Delta + 4\cos^4\Delta), \tag{III.14b}$$
$$f_2(\Delta) = 2(1 - \cos^4\Delta) \tag{III.14c}.$$

Spin-Lattice Relaxation in Nematic Liquid Crystals

With the help of expressions III(4a–b), III.6, III.7 and III.13, the spectral density $J_h(p\omega)$ – expression III.9 – becomes:

$$J_h(p\omega) = f_h(\Delta) \cdot S^2 \int_{-\infty}^{+\infty} e^{-ip\omega t} \cdot dt \cdot \sum_{\vec{q}} \sum_{\alpha=1}^{2} \langle |N_\alpha(\vec{q})|^2 \rangle \cdot e^{-t/\tau_\alpha(\vec{q})} \quad \text{(III.15)}$$

Replacing the summation over $\vec{q}$ by an integration from zero up to a cut-off value $q_{zc} = \pi/d$ and $q_{\perp c} = \pi/a$, where d and a are of the order of the molecular length and width, we find:

$$J_h(p\omega) = f_h(\Delta) \cdot S^2 \frac{1}{(2\pi)^3} \sum_\alpha \int_0^{q_{zc}} \int_0^{q_{\perp c}} \frac{kT}{(K_3 q_z^2 + K_\alpha q_\perp^2)}$$

$$\cdot \frac{2\tau_\alpha(\vec{q})}{1+p^2\omega^2\tau_\alpha(\vec{q})^2} \, 2\pi \cdot q_\perp \, dq_\perp \, dq_z$$

$$= \sum_{\alpha=1}^{2} f_h(\Delta) S^2 \frac{kT}{(2\pi)^2 K_\alpha} \frac{1}{p\omega} \frac{1}{B_\alpha} \left\{ 2B_\alpha \arctan(A_\alpha^2 + B_\alpha^2) \right.$$

$$- \frac{1}{\sqrt{2A_\alpha^2(\sqrt{1+A_\alpha^{-4}} - 1)}} \ln \frac{B_\alpha^2 - B_\alpha\sqrt{2A_\alpha^2(\sqrt{1+A_\alpha^{-4}} - 1)} + A_\alpha^2\sqrt{1+A_\alpha^{-4}}}{B_\alpha^2 + B_\alpha\sqrt{2A_\alpha^2(\sqrt{1+A_\alpha^{-4}} - 1)} + A_\alpha^2\sqrt{1+A_\alpha^{-4}}}$$

$$- \frac{2}{\sqrt{2A_\alpha^2(\sqrt{1+A_\alpha^{-4}} + 1)}} \left(\arctan \frac{2B_\alpha + \sqrt{2A_\alpha^2(\sqrt{1+A_\alpha^{-4}} - 1)}}{\sqrt{2A_\alpha^2(\sqrt{1+A_\alpha^{-4}} + 1)}} \right.$$

$$+ \arctan \left. \frac{2B_\alpha - \sqrt{2A_\alpha^2(\sqrt{1+A_\alpha^{-4}} - 1)}}{\sqrt{2A_\alpha^2(\sqrt{1+A_\alpha^{-4}} + 1)}} \right) - 2B_\alpha \arctan B_\alpha^2$$

$$- \frac{1}{\sqrt{2}} \ln \frac{B_\alpha^2 + \sqrt{2}B_\alpha + 1}{B_\alpha^2 - \sqrt{2}B_\alpha + 1} + \sqrt{2} \left[\arctan \frac{\sqrt{2}B_\alpha}{1 - B_\alpha^2} + \pi\theta(B_\alpha - 1) \right] \right\} \quad \text{(III.16)}$$

with

$$\theta = \begin{cases} 1, & B > 1 \\ 0, & B < 1 \end{cases}$$

and where

$$A_\alpha^2 = \frac{K_\alpha \cdot q_{\perp c}^2}{p \cdot \omega \cdot \eta_\alpha} = \frac{\omega_{\perp \alpha c}}{p\omega}, \quad \text{(III.17a)}$$

$$B_\alpha^2 = \frac{K_3 \cdot q_{zc}^2}{p\omega \eta_\alpha} = \frac{\omega_{z\alpha c}}{p\omega}. \quad \text{(III.17b)}$$

In deriving (III.16) the q-dependence of the viscosities [1] η_α was neglected. A subsequent numerical integration of expressions (III.15), including the q-dependence of η_α, differs by less than 10 % from the analytical form (III.16). The above approximation thus does seem to be quite good.

Inserting (III.16) into (III.18), the Zeeman spin-lattice relaxation rate T_1^{-1} is obtained as a function of the Larmor frequency ω and the angle Δ between $\vec{N}_0$ and $\vec{H}_0$.

It should be noted that if the system is perfectly aligned along $\vec{H}_0$ so that $\Delta = 0$, then only the term with h = 1 is different from zero. This is so as $f_1(\Delta = 0) = 1$, $f_0(\Delta = 0) = f_2(\Delta = 0) = 0$. In such a case

$$T_1^{-1} = \frac{2}{5} T_{1\rho}^{-1} = \frac{9}{8} \gamma^4 \hbar^2 r^{-6} J_1(\omega_0). \tag{III.18}$$

In the general case however both $J_1(\omega_0)$ and $J_2(2\omega_0)$ will contribute to T_1^{-1}, whereas $T_{1\rho}^{-1}$ will in addition to $J_1(\omega_0)$ and $J_2(2\omega_0)$ "feel" the contribution of $J_0(2\omega_1)$.

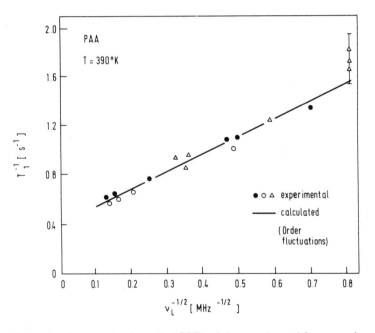

Fig. 1. Comparison between the theoretical ODF and the experimental frequency dependence [5] of the proton spin-lattice relaxation time in nematic PAA

Let us now look at the frequency dependence of T_1^{-1} as given by (III.16). We have to discuss several different limiting cases:

(i) $A_\alpha, B_\alpha \gg 1$, i.e. $\omega \ll \omega_{l\alpha c}, \omega_{z\alpha c}$: In this case expression (III.16) simplifies to

$$J_h(p\omega) = \sum_{\alpha=1}^{2} \frac{C' \pi \sqrt{2}}{2 K_\alpha} \cdot \sqrt{\frac{\eta_\alpha}{K_3}} \cdot \frac{1}{\sqrt{p\omega}} \tag{III.19a}$$

with

$$C' = f_h(\Delta) \cdot S^2 \cdot \frac{kT}{2\pi^2}.$$ (III.19b).

This limit thus gives the same frequency dependence of T_1

$$T_1 \propto \sqrt{\omega}$$

as obtained in the one elastic constant ($K_1 = K_2 = K_3$) approximation [2] and experimentally verified in PAA [5] (Fig. 1). As

$$\omega_{\perp 1c} = \frac{K_1 \cdot q_{1c}^2}{\eta_1} \approx 10^3 \times 2\pi \text{ MHz}$$ (III.19c)

with $\ell \approx 10$ Å, $K_1 \sim 6 \times 10^{-7}$ dyn, $\eta_1 \approx 0.1$ Poise, and $\omega_{\perp 2c}$, ω_{zac} are usually within a factor of ten equal to $\omega_{\perp 1c}$, the nuclear Larmor frequencies will be in most NMR experiments nearly always much smaller than the "cut-off" frequencies: $\omega \ll \omega_c$. The above limit — which physically means that the continuum theory can be used for the whole range of q-values and that the integration over q in (III.16) can be performed from zero to ∞ — will thus apply to most T_1^{-1} experiments in the nematic phase.

(ii) $A_\alpha, B_\alpha \ll 1$, i.e. $\omega \gg \omega_{\perp ac}, \omega_{zac}$: In this limit, which is just the opposite to the case discussed above, we find:

$$J_h(p\omega) \approx C' \cdot \frac{q_{1c}^2 \cdot q_{zc}^2}{2p^2\omega^2} \cdot \left(\frac{1}{\eta_1} + \frac{1}{\eta_2}\right) \propto \frac{1}{\omega^2}$$ (III.20)

so that $T_1 \propto \omega^2$. However, as already mentioned, this case will be hardly ever realized in nematic liquid crystals. The same is true for

(iii) $A_\alpha \ll 1, B_\alpha \gg 1$, i.e. $\omega_{zac} \gg \omega \gg \omega_{\perp ac}$: where

$$J_h(p\omega) = C' \frac{q_{1c}^2}{2K_3^2 q_{zc}^3} (\eta_1 + \eta_2) \neq f(\omega)$$ (III.21)

and

(iv) $A_\alpha \gg 1, B_\alpha \ll 1$, i.e. $\omega_{\perp ac} \gg \omega \gg \omega_{zac}$: where

$$J_h(p\omega) = \frac{C' \cdot \pi \cdot q_{zc}}{4 \cdot K_1 \cdot p \cdot \omega} \propto \frac{1}{\omega}.$$ (III.22).

It should be finally mentioned that in the one elastic constant approximation ($K_1 = K_2 = K_3$) the expression for T_1^{-1} derived in Ref. [4] is recovered:

$$T_1^{-1} = C \cdot \frac{1}{\sqrt{\omega}} \left[f_1(\Delta) g(a) + \frac{1}{\sqrt{2}} \cdot f_2(\Delta) \cdot g(\sqrt{2}\,a) \right]$$ (III.23)

where

$$g(\sqrt{pa}) = \pi - \frac{1}{2} \ln \frac{pa^2 + \sqrt{2pa} + 1}{pa^2 - \sqrt{2pa} + 1} - \mathrm{arctg}(\sqrt{2p} \cdot a + 1) - \mathrm{arctg}(\sqrt{2p} \cdot a - 1)$$
(III.24)

and $a = \sqrt{\dfrac{\omega}{\omega_c}}$ with $\omega_c = Kq^2/\eta$, whereas $C = \dfrac{9\sqrt{2}\gamma^4 \hbar^2 \cdot S^2 \cdot kT}{16\pi^2 \cdot r^6 \cdot K} \sqrt{\dfrac{\eta}{K}}$.

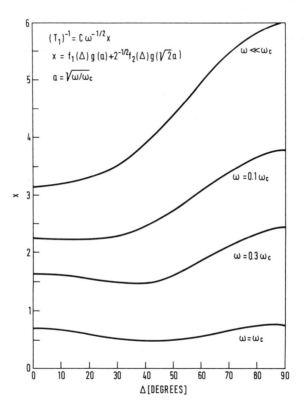

Fig. 2. Theoretical angular dependence [9] of the proton T_1^{-1} for the nematic order director fluctuation (ODF) mechanism for different ratios of ω/ω_c. Here Δ is the angle between the nematic director and the magnetic field direction.

The angular dependence of T_1^{-1} — according to expression (III.23) — is shown in Fig. 2. for different ratios ω/ω_c. It can be seen that for $\omega \ll \omega_c$, T_1^{-1} increases for nearly a factor of two between $\Delta = 0°$ and $\Delta = 90°$, whereas T_1^{-1} practically does not depend on Δ for $\omega = \omega_c$. Unfortunately there are no experimental data with which one could compare the predicted angular dependence of T_1^{-1}. The measured $T_{1\rho}^{-1}$ data in nematic MBBA–EBBA mixtures (Fig. 3), on the other hand, agree with the angular dependence given by expression (III.11):

$$T_{1\rho}^{-1} \propto \cos^2\Delta - \cos^4\Delta$$

Spin-Lattice Relaxation in Nematic Liquid Crystals

which predicts a minimum in $T_{1\rho}$ around $\Delta = 60°$. The angular dependence of the inverse dipolar splitting, which also agrees with the theoretical one $- \delta\alpha(3\cos^2\Delta - 1)^{-1} -$ is shown in Fig. 4.

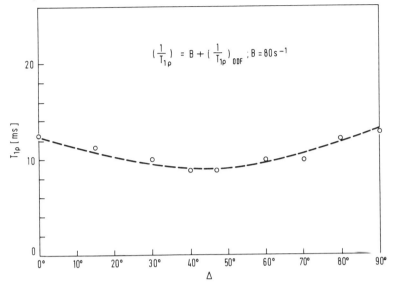

Fig. 3. Comparison between the experimental and the theoretical ODF angular dependence of the proton spin-lattice relaxation rate in the rotating frame, $T_{1\rho}^{-1}$. The data were taken for a MBBA-EBBA nematic mixture at $-20\ °C$ by turning the sample in the magnetic field for a preset angle. This technique exploits the fact that the relaxation time of the nematic director in a magnetic field is in this system much longer than $T_{1\rho}$

The temperature dependence of T_1^{-1} is given by $\dfrac{S^2 \cdot T}{K}\sqrt{\dfrac{\eta}{K}}$ and should be weak ($K \propto S^2$) as it has been indeed observed [2, 3].

It should be noted that the $T_1^{-1} \propto \omega_L^{-1/2}$ dispersion law as given by expression (III.19a) can not hold at arbitrarily low Larmor frequencies where it predicts an infinite relaxation rate, $T_1^{-1} \to \infty$ as $\omega_L \to 0$, so that the ODF mechanism would become dominant in any nematic system at low enough frequencies. This singularity disappears if one takes into account that the nematic order correlation length ξ is finite due to the presence of disclinations and other defects. In such a case only order fluctuations with a wavelength smaller than ξ and larger than the molecular dimensions ℓ can take place and we find — by a straightforward extension of the Pincus [2] and Doane [4] calculations — the spectral density of the angular part F_h of the dipole-dipole interactions modulated by ODF as:

$$J_h(p\omega_L) = \int_{-\infty}^{+\infty} \langle F_h(0)F_h^*(t)\rangle e^{-i\omega t}\,dt = D \cdot \int_{q_{min}}^{q_{max}} \dfrac{q^2\,dq}{q^4 + p^2\omega_L^2\eta^2/K^2}$$

$$= \dfrac{C_1}{\sqrt{p\omega_L}} \cdot [g(a_{max}) - g(a_{min})] \qquad (III.25)$$

where

$$g(a) = 2 \arctg \frac{a\sqrt{2}}{a^2 - 1} - \ln \frac{1 + a\sqrt{2} + a^2}{1 - a\sqrt{2} + a^2}$$

$$a_{max} = \sqrt{\frac{p\omega_L}{\omega_{c,max}}} = \sqrt{\frac{p\omega_L \eta}{K q_{max}^2}}, \quad a_{min} = \sqrt{\frac{p\omega_L}{\omega_{c,min}}} = \sqrt{\frac{p\omega_L \eta}{K q_{min}^2}}$$

$$q_{max} = \frac{2\pi}{\ell}, \quad q_{min} = 2\pi/\xi,$$

$$C_1 = \frac{kT S^2 \sqrt{\eta} f_h(\Delta)}{2\sqrt{2}\pi K \sqrt{K}}, \quad T_1^{-1} = \text{const}\,[J_1(\omega_L) + J_2(2\omega_L)]$$

and where the notation follows that of Ref. [6].

The above expressions show that the nematic ODF relaxation mechanism becomes ineffective when $\omega_L \gg \omega_{c,max}$ and that T_1^{-1} approaches a constant value for $\omega_L \ll \omega_{c,min}$. Only for $\omega_{c,min} \ll \omega_L \ll \omega_{c,max}$ the "$\omega_L^{1/2}$" dispersion law holds.

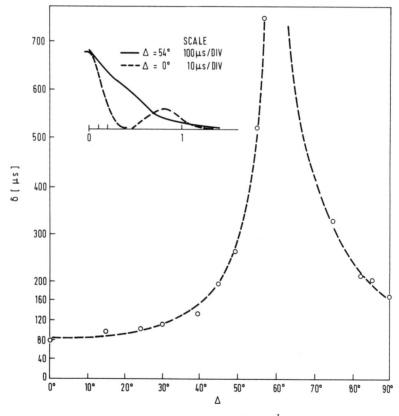

Fig. 4. Angular dependence of the inverse dipolar splitting $\delta = \frac{1}{\gamma \Delta H}$ of the proton line in nematic MBBA-EBBA mixtures using the same technique as in Fig. 3

References

[1] de Gennes, P. G.: The Physics of Liquid Crystals. Oxford: Clarendon Press (1974) and references therein.
[2] Pincus, P.: Solid State Commun. 7, 415 (1969).
[3] Blinc, R., Hogenboom, D. L., O'Reilly, D. E., Peterson, E. M.: Phys. Rev. Letters 23, 969 (1969).
[4] Doane, J. W., Johnson, D. L.: Chem. Phys. Letters 6, 291 (1970).
Doane, J. W., Tarr, C. E., Nickerson, M. A.: Phys. Rev. Letters 33, 622 (1974).
[5] Vilfan, M., Blinc, R., Doane, J. W.: Solid State Commun. 11, 1073 (1972).
[6] Blinc, R., Lugomer, S. Žekš, B.: Phys. Rev. A9, 2214 (1974).
[7] Cabane, B., Clark, W.: Phys. Rev. Letters 25, 91 (1970).
[8] McElroy, R., Thompson, R., Pintar, M.: Phys. Rev. A10, 403 (1974).
[9] Blinc, R., Luzar, M., Vilfan, M., Burgar, M.: J. Chem. Phys. 63, 3445 (1975).

NMR Studies of Molecular Tunnelling

S. Clough

Contents

1. Interpretation of Magnetic Resonance Experiments Involving Tunnelling Groups ... 113
   1.1 Hopping Motion of Methyl Group ... 114
   1.2 Quantum Rotation of Methyl Groups ... 115
   1.3 The Spin Hamiltonian ... 116
   1.4 Combined Tunnelling and Hopping Rotation ... 116
   1.5 Symmetry Adaptation of $\mathcal{H}_s$ ... 117
   1.6 NMR Line Width for Rotating Methyl Group ... 118
2. Coupled Nuclear and Rotational Relaxation ... 120
   2.1 Relaxation of Tunnelling Methyl Groups ... 121
   2.2 The Haupt Effect ... 122
3. References ... 123

1. Interpretation of Magnetic Resonance Experiments Involving Tunnelling Groups

The motion of atoms and molecules in solids has the effect of modulating the magnetic dipole-dipole interactions between nuclear spins or between nuclear spins and unpaired electron spins, thereby affecting magnetic resonance line-shapes and relaxation times. Through these experimentally observable effects, detailed information on the molecular motion can often be inferred.

The great majority of magnetic resonance studies of molecular motion have been conducted at relatively high temperatures (above 70 K). These results can be interpreted in terms of a simple classical model of the motion. For a long time however it has been realized that quantum mechanical features of the motion become important at low temperatures. Experimental evidence for this first emerged in the context of the traditional NMR measurements of line width (or T_2) and spin lattice relaxation time T_1, since in some materials, especially those containing small symmetric groups, the temperature dependence of these parameters could not be reconciled with the classical model of molecular motion. More recently, the coherent character of quantum motions has led to new types of measurements whose object is to measure a motional frequency rather than a rate constant. From such measurements emerges a description of molecular motion which spans both low and high temperatures and includes both coherent and random motions.

1.1 Hopping Motion of Methyl Group

The usual model for the hindered rotation of methyl groups in solids is of a rigid equilateral triangle confined by a hindering barrier to small oscillations about its symmetry axis. Rotation of the group is visualized as a sudden transition from one orientation to another equivalent orientation. This avoids a detailed consideration of the mechanism of the hopping process, though it clearly requires that the group absorb energy from the crystal lattice in order to surmount the hindering barrier. This classical concept finds a parallel even when quantum mechanics is used.

As a result of the geometrical symmetry of the group, the potential energy has threefold symmetry as a function of the rotational coordinate ϕ and may be written as a Fourier series.

$$V(\phi) = \frac{1}{2} \sum_n V_n (1 - \cos 3n\phi). \tag{1}$$

This can be expanded around the three potential minima to give

$$V(\phi) \cong \frac{9}{4} (\phi - \phi_0)^2 \sum_n n^2 V_n = \frac{1}{2} I\omega^2 (\phi - \phi_0)^2 \tag{2}$$

with $\phi_0 = 0$ or $\pm 2\pi/3$. I is the moment of inertia of the group. From (2) we have the harmonic oscillator Hamiltonian

$$\mathcal{H}_r = -(\hbar^2/2I)\partial^2/\partial\phi^2 + \frac{1}{2} I\omega^2 (\phi - \phi_0)^2 \tag{3}$$

whose eigenvalues are $E_k = \hbar\omega(k + \frac{1}{2})$. The two lowest states are

$$\psi_0 = (I\omega/\hbar\pi)^{\frac{1}{4}} \exp(-I\omega(\phi - \phi_0)^2/2\hbar)$$
$$\psi_1 = (2\pi)^{\frac{1}{2}} (I\omega/\hbar\pi)^{\frac{3}{4}} (\phi - \phi_0) \exp(-I\omega(\phi - \phi_0)^2/2\hbar). \tag{4}$$

The spacing of the harmonic oscillator levels $\hbar\omega$ is determined by the barrier through (2). If as is commonly assumed, $V_n = 0$ except for $n = 1$, then $\omega = \sqrt{(9V/2I)}$. In most molecular solids $\hbar\omega/k$ exceeds 50 K, so that at low temperatures, most of the methyl groups are in the ground torsional oscillator state.

Hopping rotation may be considered as a two step process. First the methyl group is excited to a high energy state which is not a harmonic oscillator state because the approximation (2) is only valid for the lowest levels; then the methyl group falls back to a ground state harmonic oscillator state, but in a different potential well from which it originated. The process is very complicated since it involves changes in the state of the lattice (scattering of phonons) as well as rotation of the methyl group. It is this fact combined with our ignorance of the state of the lattice (except in a statistical sense), which is the justification for regarding the process as occurring randomly in time. Phenomenologically it may then be described in terms of a rate constant τ^{-1}, the rotational transition itself being regarded as instantaneous.

It is often found that the temperature dependence of τ^{-1} is given by

$$\tau^{-1} = \tau_0^{-1} \exp(-E/kT) \tag{5}$$

a fact which indicates that the two step rotational process occurs mainly via levels at about E above the ground torsional oscillator state. There is no need to consider that the methyl group must be excited to a level above the top of the hindering barrier before rotation can occur, since this would be to use a classical mechanics we know to be wrong, and the quantum mechanical wavefunctions just above and just below the top of the barrier are not fundamentally different. Nonetheless, the activation energy appears to be comparable with the height of the hindering barrier above the ground torsional state.

1.2 Quantum Rotation of Methyl Groups

The functions (4) are not eigenfunctions when $V(\phi)$ is used instead of the harmonic approximation of (3). The most important consequence of this replacement from our point of view is to mix harmonic oscillator states of different potential wells. The Hamiltonian is

$$\mathcal{H}'_r = -(\hbar^2/2I)\partial^2/\partial\phi^2 + V(\phi) \tag{6}$$

and we take account of the off-diagonal elements

$$\int_0^{2\pi} \psi_0(\phi) \mathcal{H}'_r \psi_0(\phi + 2\pi/3) d\phi = -h\nu_t/3 . \tag{7}$$

Linear combinations which diagonalize $\mathcal{H}'_r$ are given by

$$\psi_0^\lambda = (1/3)^{\frac{1}{2}} (\psi_0(\phi) + \lambda\psi_0(\phi - 2\pi/3) + \lambda^*\psi_0(\phi + 2\pi/3)) \tag{8}$$

where $\lambda = 1$ or ϵ or ϵ^* with $\epsilon = \exp(2\pi i/3)$. The energy levels are given by E^λ with

$$E^1 = E_0 - 2h\nu_t/3 \; ; E^\epsilon = E^{\epsilon^*} = E_0 + h\nu_t/3 . \tag{9}$$

E_0 of course is close to the harmonic approximation of $\frac{1}{2}\hbar\omega$, but the important feature is the splitting of the previously degenerate ground state into a singlet and a doublet. That this is connected with tunnelling rotation of the group can be seen, not only from the form of (7) but more explicitly by considering the evolution of the state $\psi_0(t)$, which at $t = 0$ is $\psi_0(\phi)$. This state is a linear combination of the stationary states (8) to which we attach the time dependence derived from (9).

$$\psi_0(t) = 3^{-\frac{1}{2}} (\psi_0^1 + (\psi_0^\epsilon + \psi_0^{\epsilon^*}) \exp(2\pi i\nu_t t)). \tag{10}$$

At $t = 1/2\nu_t$, this becomes

$$3^{-1}(-\psi_0(\phi) + 2\psi_0(\phi + 2\pi/3) + 2\psi_0(\phi - 2\pi/3))$$

which is weighted 1/9 : 4/9 : 4/9 in the three orientations. Thus an initially localized orientation oscillates between that orientation and a combination involving the other tow, with a frequency ν_t.

The first excited torsional oscillator state exhibits an analogous tunnelling splitting $h\nu_t^1$, given in the harmonic approximation by the analogue of (7). The integral has the opposite sign from (7), because $\psi_1(\phi)$ is an odd function while $\psi_0(\phi)$ is even. The tunnelling splitting of the first excited state is therefore inverted compared with the ground state splitting.

Though we have discussed the torsional oscillator tunnelling states (7) in terms of a harmonic approximation, there is no difficulty in computing exact eigenvalues and eigenfunctions of $\mathcal{H}_r'$. Levels of the potential $V(\phi) = \frac{1}{2} V(1 - \cos 3\phi)$ are a function of the barrier height V. For $V = 0$ the problem is trivial. The eigenfunctions are the free rotor functions $\exp(\pm im\phi)$ with $m = 0$ or integral. The eigenvalues are $\hbar^2 m^2/2I$.

The pattern of singlets and doublets is not dependent on our assumption of a rigid triangular group with threefold geometrical symmetry. It arises in a very fundamental way due to the symmetry of the Hamiltonian under cyclic permutation of the coordinates of the three protons, and depends only on the indistinguishability of the protons.

1.3 The Spin Hamiltonian

It is usual to treat magnetic resonance problems using a Hamiltonian in which the only operators to appear are spin operators, spatial coordinates occurring only as classical variables. A quantum mechanical description of molecular motion is arrived at with a Hamiltonian containing operators with both spatial and spin variables. Since the only feature we wish to introduce is the splitting of the torsional ground state, a much simpler option is available.

First we may define an operator which cyclically permutes the space coordinates of the three methyl group protons.

$$R\psi_0(\phi) = \psi_0(\phi + 2\pi/3) = R^{-1}\psi_0(\phi - 2\pi/3). \tag{11}$$

Then so long as we limit ourselves to the ground torsional state, we may replace $\mathcal{H}_r'$ by $\mathcal{H}_t$ since both have the same eigenvalues with respect to (8) except for an unimportant additive constant.

$$\mathcal{H}_t = -(h\nu_t/3)(R + R^{-1}). \tag{12}$$

We may similarly define an operator P which cyclically permutes the spin coordinates of the protons. The indistinguishability of the protons means that PR is the unit operator. Thus (12) may be replaced by the spin operator

$$\mathcal{H}_t = -(h\nu_t/3)(P + P^{-1}). \tag{13}$$

The effect of tunnelling rotation is then easily handled by adding (13) to the spin Hamiltonian.

1.4 Combined Tunnelling and Hopping Rotation

While the effect of tunnelling rotation is absorbed into the Hamiltonian through (13), hopping rotation must be treated differently since it has the character of transitions between states. A hopping transition has the effect of converting a state ψ into $R\psi$.

As we have seen, we cannot distinguish between cyclic permutations of space coordinates and the reverse cyclic permutation of spin coordinates. Once again therefore, we prefer to regard a hop as a sudden cyclic permutation of spin coordinates, in order to simplify the analysis. The effect of the hopping transitions may be put into a convenient form by considering an ensemble of systems, each of which is described at some instant of time by a state function ψ

$$\psi = \sum_j a_j u_j \tag{14}$$

which is expanded in a set of basis states u_j. The coefficients a_j are dependent on time due to the hopping process. During a short interval dt a fraction dt/τ hop clockwise and are converted to the state $P\psi$ and a similar fraction are converted to $P^{-1}\psi$. Each of these new state functions can be expanded as in (14), though with different coefficients. The coefficient which was initially a_k is converted to either $\sum_j a_j P_{kj}$ or $\sum_j a_j P_{kj}^{-1}$ where $P_{kj} = \langle u_k | P | u_j \rangle$.

This change in a_k results in a corresponding change in $a_k^* a_m$ whose average value is the density matrix element ρ_{mk}. This change is $(\sum_i \sum_j a_i^* a_j P_{ki}^* P_{mj}) - a_k^* a_m$ for a clockwise rotation occurring with a probability dt/τ. Including a similar term for anticlockwise rotations and averaging over the ensemble one obtains

$$d\rho_{mk} = (dt/\tau)(\sum_i \sum_j \rho_{ji}(P_{ki}^* P_{mj} + P_{ki}^{-1*} P_{mj}^{-1}) - 2\rho_{mk}).$$

This is easily rearranged, using the fact that $P_{ki}^* = P_{ik}^{-1}$, to give

$$d\rho/dt = (P\rho P^{-1} + P^{-1}\rho P - 2\rho)/\tau \tag{15}$$

which describes the evolution of the density matrix due to the hopping process [1]. Then the full differential equation for the density matrix is given by

$$d\rho/dt = i\hbar^{-1}[\rho, (\mathcal{H}_s + \mathcal{H}_t)] + (P\rho P^{-1} + P^{-1}\rho P - 2\rho)/\tau. \tag{16}$$

1.5 Symmetry Adaptation of $\mathcal{H}_s$

Before attempting to analyse a particular problem with the help of (16), it is of considerable assistance to rearrange $\mathcal{H}_s$ into terms which belong to the irreducible representations of the permutation group consisting of P, P^{-1} and the identity operator. This is best illustrated by an example. For a single methyl group, $\mathcal{H}_s$ consists of the Zeeman term and the inter-proton dipole-dipole interactions.

$$\mathcal{H}_s = -\hbar\omega_0 \sum_j I_{jz} + \sum_{j>k} \left(\sum_{m=-2}^{m=2} B_{jk}^m F_{jk}^{-m} \right). \tag{17}$$

The suffices $j, k = 1, 2, 3$ label the protons at fixed sites. The dipole-dipole interaction between each pair of protons is separated as usual into five terms each of which is a product between a space function B and a proton spin operator F. The terms with $m = 0$ commute with the Zeeman term and determine the line width while the remaining terms influence the spin lattice relaxation. Now we rewrite the dipole-dipole terms of (17) is follows

$$\sum_{j>k} B_{jk}^m F_{jk}^{-m} = \sum_\lambda B_\lambda^m F_{\lambda*}^{-m} \tag{18}$$

where as in (8), λ takes the values 1, ϵ and ϵ^* so that there are three terms on each side of (18). The terms on the right are defined through

$$B_\lambda^m = 3^{-\frac{1}{2}} (B_{12}^m + \lambda B_{23}^m + \lambda^* B_{31}^m) \tag{19}$$

With this definition

$$PF_1 = F_1 P \,; PF_\epsilon^{-m} = \epsilon F_\epsilon^{-m} P \,; PF_{\epsilon*}^{-m} = \epsilon^* F_{\epsilon*}^{-m} P \,. \tag{20}$$

The relations (20) enormously simplify the evaluation of matrix elements.

1.6 NMR Line Width For Rotating Methyl Group

The case of no rotation follows from putting $\nu_t = 0$ and $1/\tau = 0$ in (16). The NMR line shape is then determined by the secular $m = 0$ terms of (17). The case of rapid hopping motion can be dealt with as follows [2]. Formally, we may transform away the hopping motion so that (16) becomes

$$d\rho'/dt = i\hbar^{-1}[\rho', (\mathcal{H}'_s + \mathcal{H}'_t)] \tag{21}$$

and

$$d\mathcal{H}'_s/dt = (P\mathcal{H}_s P^{-1} + P^{-1}\mathcal{H}_s P - 2\mathcal{H}_s)/\tau \tag{22}$$

it being understood that $\mathcal{H}'_s$ is a random operator and (22) describes an ensemble average. The symmetry adapted terms of (18) now exhibit simple results when inserted into (22) since

$$PF_1 P^{-1} = F_1 \,; \; PF_\epsilon P^{-1} = \epsilon F_\epsilon \,; PF_{\epsilon*} P^{-1} = \epsilon^* F_{\epsilon*} \,. \tag{23}$$

Thus the terms F'_1 are time independent, while the terms F'_ϵ and $F'_{\epsilon*}$ fluctuate with the average time dependence

$$dF'_\epsilon/dt = -3F'_\epsilon/\tau \tag{24}$$

giving a correlation function for the fluctuating operator $F'(t)$

$$F'_\epsilon(0) F'_{\epsilon*}(t) = F'_\epsilon(0) F'_{\epsilon*}(0) \exp(-3t/\tau) \,. \tag{25}$$

If the hopping rate is rapid then the fluctuating terms in $\mathcal{H}_s$ have little influence on the evolution of ρ' according to (21) except on the time scale of τ. They may then be dropped from (31) and the subsequent line shape calculation. Thus the line width is now dependent only on those dipole-dipole terms in (17) which have $m = 0$ and $\lambda = 1$ and the line is narrowed.

Next we may consider the case of motional narrowing due to tunnelling rotation only [3], by putting $1/\tau = 0$. This time we transform out of $\mathcal{H}_t$. Instead of (21) we have

$$d\rho'/dt = i\hbar^{-1} [\rho, \mathcal{H}'_s] \tag{26}$$

with $\mathcal{H}'_s = \exp(i\hbar^{-1}\mathcal{H}_t t) \mathcal{H}_s \exp(-i\hbar^{-1}\mathcal{H}_t t)$. The resulting time dependence of the term in $\mathcal{H}'_s$ deriving from F_λ is given by

$$dF'_\lambda/dt = i\hbar^{-1}[\mathcal{H}_t, F_\lambda]. \tag{27}$$

Once again the terms with $\lambda = 1$ are time independent while the remaining terms oscillate with frequency ν_t. If this frequency is sufficiently high, then the oscillating terms may be dropped from $\mathcal{H}'_s$. Thus, to this approximation the motionally narrowed spectrum due to tunnelling rotation is the same as that when narrowing is due to hopping rotation.

The difference is in the character of the modulation of the time dependent terms which fluctuate randomly in the case of hopping rotation and oscillate coherently with tunnelling rotation. The difference in the predicted spectrum must be sought in the wings of the spectrum. For hopping rotation, broad wings are predicted extending to a frequency $(2\pi\tau)^{-1}$ on each side of the spectrum. For tunnelling rotation, discrete lines are predicted at a frequency of the order of ν_t on each side of the main spectrum centred at the nuclear Larmor frequency. Detection of these tunnelling sidebands is one way of measuring ν_t, but the intensity of these sidebands is usually too weak to be detected.

The NMR spectra of powders containing methyl groups have been calculated for various values of ν_t with $1/\tau = 0$ and for values of $1/\tau$ for $\nu_t = 0$. The general problem would be more laborious, and is perhaps of limited interest since normally one of the rotational processes is dominant. The procedure though is straightforward.

Eq. (16) can be written

$$d\rho_{ij}/dt = -\sum_{mn} A_{ij,mn}\rho_{mn}. \tag{28}$$

The elements of ρ may be regarded as the components of a vector and $A_{ij,mn}$ as a matrix. This can be diagonalized by a transform S.

$$A' = SAS^{-1}; \rho' = S\rho; d\rho'_{pq}/dt = -A'_{pq,pq}\rho'_{pq}. \tag{29}$$

After a 90° pulse the density matrix has the form

$$\rho(0) = (1 + aI^y)/\text{Tr}(1) \tag{30}$$

where $I^y = I_{1y} + I_{2y} + I_{3y}$. Integration of (29) gives the density matrix at a later time

$$\rho_{uv}(t) = \sum_{pq} S^{-1}_{uv,pq}\rho'_{pq}(t),$$

$$\rho'_{pq}(t) = \sum_{mn} S_{pq,mn}(1 + aI^y)_{mn}\exp(-A'_{pq,pq}t)/\text{Tr}(1),$$

and the expectation value of I^y is $\text{Tr}(\rho I^y)$

$$\langle I^y \rangle = (a/\text{Tr}(1))\sum_{uv}\sum_{pq}\sum_{mn} I^y_{uv} S^{-1}_{uv,pq} S_{pq,mn} I^y_{mn} \exp(-A'_{pq,pq}t).$$

This gives the free induction decay and its Fourier transform is the line shape. The solution is seen to depend on the diagonalization of a large complex matrix for which standard computer programmes are available.

2. Coupled Nuclear and Rotational Relaxation

In analysing relaxation phenomena in magnetic resonance, one normally has in mind a set of energy levels whose populations, having been disturbed from the thermal equilibrium values, are returning to these values due to transitions between the levels. Often the disturbance is achieved by means of pulses of radiofrequency field which transfer populations between spin levels. Alternatively, a disturbance can be achieved simply by changing the sample temperature which changes the equilibrium values. This has the advantage that it applies to rotational levels as well as spin levels and for this reason is the chief method discussed here.

If only transitions between single pairs of levels are considered, then the rate equations for the deviations of the populations from thermal equilibrium values are

$$dp_i/dt = \sum_j W_{ij}(p_j - p_i) \qquad (31)$$

where W_{ij} is the transition probability between states i and j. If one also takes account of flip-flop transitions in which an upward transition between one pair of levels is accompanied by a downward transition between a second pair of levels, then these introduce terms with the form

$$dp_i/dt = \sum_j \sum_{u,v} W_{ijuv}((P_j + p_j)(P_v + p_v) - (P_i + p_i)(P_u + p_u)) \qquad (32)$$

where P_i is the thermal equilibrium value of the population of the ith level. The transition probability is zero unless energy is conserved, which means $P_j P_v = P_i P_u$. The high temperature approximation implies that $p_i \ll P_i$ so that (32) is well approximated by a set of linear equations

$$dp_i/dt = -\sum_j G_{ij} p_j \qquad (33)$$

which may also be taken to include higher order processes. The relative magnitudes of the transition probabilities W_{ij} which describe the way the system equilibrates with the lattice and the G_{ij} which determine the rate the system approaches an internal quasi-equilibrium, govern the way that the problem is analysed.

If the rate constants G_{ij} are most important, as is the case for the nuclear flip-flop terms in the dipole-dipole interactions in solids, then we may diagonalize (33) to obtain

$$dx_i/dt = -R_i x_i, \qquad (34)$$

$$x_i = \sum_j S_{ij} p_j, \qquad (35)$$

where S_{ij} is an orthogonal matrix. If $R_i \neq 0$, then (34) is a good enough approximation; if $R_i = 0$ then terms from (31) need to be introduced. In the latter case the x_i are quasi-invariants, changing only on the time scale of the W_{ij} rather than on the time scale G_{ij}; i.e., at a rate of the order T_1^{-1} rather than T_2^{-1}.

Two such quasi-invariants can be identified from a simple property of the transition probabilities W_{ijuv}. There are zero unless the transition conserves both the Zeeman and dipolar energies. It follows that the combinations x_Z, x_D satisfy (34) with $R = 0$.

$$x_Z = \sum_j Z_j p_j / (\sum_j Z_j^2)^{\frac{1}{2}}, \tag{36}$$

$$x_D = \sum_j D_j p_j / (\sum_j D_j^2)^{\frac{1}{2}}, \tag{37}$$

where Z_j and D_j are the Zeeman and dipolar energies in state j. Where there is no molecular motion it is assumed that these are the only quasi-invariants of the spin system.

If at $t = 0$, the spin system is disturbed from thermal equilibrium so that the x_i are non-zero, then after a short time, only the quasi-invariants are non-zero, the remainder having decayed according to (34). The p_i are then given from (35)

$$p_i = (Z_i x_Z / (\sum_j Z_j^2)^{\frac{1}{2}}) + (D_i x_D / (\sum_j D_j^2)^{\frac{1}{2}}). \tag{38}$$

Differential equations describing the evolution of x_Z and x_D under the influence of spin lattice relaxation are now obtained by differentiating (36) and (37) and substituting (31) and (38).

$$dx_Z/dt = -\frac{1}{2} x_Z \sum_{i,j} W_{ij}(Z_j - Z_i)^2 / (\sum_j Z_j^2), \tag{39}$$

$$dx_D/dt = -\frac{1}{2} x_D \sum_{i,j} W_{ij}(D_j - D_i)^2 / (\sum_j D_j^2). \tag{40}$$

Thus in this case, exponential relaxations are expected for both Zeeman and dipolar energies.

2.1 Relaxation of Tunnelling Methyl Groups

The spin Hamiltonian describing a sample containing rapidly tunnelling methyl groups has the form

$$\mathcal{H} = \mathcal{H}_Z + \mathcal{H}_D + \mathcal{H}_t.$$

Each of these terms introduces its own characteristic splittings into the energy level diagram. Flip-flop transitions due to inter-proton dipole-dipole interactions do not occur unless they separately conserve Zeeman, dipolar and rotational energies. We may therefore identify a third quasi-invariant of the motion, namely

$$x_t = \sum_j T_j p_j / (\sum_j T_j^2)^{\frac{1}{2}} \tag{41}$$

where T_j is the expectation value of $\mathcal{H}_t$ in state j. The equation corresponding to (38) is now

$$p_i = (Z_i x_Z / (\sum_j Z_j^2)^{\frac{1}{2}}) + (D_i x_D / (\sum_j D_j^2)^{\frac{1}{2}}) + (T_i x_t / (\sum_j T_j^2)^{\frac{1}{2}}). \tag{42}$$

Eq. (39) is not affected, but (40) is modified as follows:

$$\begin{aligned} dx_D/dt &= -Ax_D - Cx_t \\ dx_t/dt &= -Cx_D - Bx_t \end{aligned} \tag{43}$$

where

$$A = \frac{1}{2} \sum_{i,j} W_{ij}(D_j - D_i)^2 / (\sum_j D_j^2),$$

$$B = \frac{1}{2} \sum_{i,j} W_{ij}(T_j - T_i)^2 / (\sum_j T_j^2), \qquad (44)$$

$$C = \frac{1}{2} \sum_{i,j} W_{ij}(D_j - D_i)(T_j - T_i) / (\sum_j D_j^2)^{\frac{1}{2}} (\sum_j T_j^2)^{\frac{1}{2}}.$$

The fact that x_Z is uncoupled to x_t and x_D is due to the equality between W_{ij} and W_{uv} when $m_i = -m_u$ and $m_j = -m_v$ where these are the magnetic quantum numbers. The cross coupling term then disappears, since the contributions cancel in pairs.

2.2 The Haupt Effect

The coupling between dipolar magnetization and molecular rotation was discovered by Haupt [4] in 4-methyl pyridine. The effect occurs only below about 45 °K. A sudden change in temperature is followed by the growth and subsequent decay of an enormous dipolar signal which is readily detected by applying a small rf inspection pulse.

Measurements follow the behaviour of the coupled Eqs. (43) [5]. Following the sudden change of temperature both x_D and x_t are non-zero, but the latter is much greater than the former because the splittings due to $\mathcal{H}_t$ greatly exceed those due to $\mathcal{H}_D$ by a factor of order 10^6. We may therefore regard the initial conditions following the thermal disturbance as $x_t = x_0; x_D = 0$. Integration of (43) then gives

$$x_D = Cx_0(k_1 - k_2)^{-1} (\exp(-k_1 t) - \exp(-k_2 t)) \qquad (45)$$

where

$$k_1 = \frac{1}{2}(A+B) - \frac{1}{2}((A-B)^2 + 4C^2)^{\frac{1}{2}},$$

$$k_2 = \frac{1}{2}(A+B) + \frac{1}{2}((A-B)^2 + 4C^2)^{\frac{1}{2}}, \qquad (46)$$

It will be apparent from (45) that if C is comparable with $A - B$, then x_D achieves a maximum value comparable to x_0. In effect, the population differences associated with the big splittings due to $\mathcal{H}_t$ have been transferred to levels split only by the small splittings of $\mathcal{H}_D$.

The validity of (43) as a description of the system may be demonstrated by a simple variation of the experiment which involves applying a radiofrequency pulse, or series of pulses at a time t_0 following the sudden temperature change. This has the effect of making $x_D = 0$ at t_0, so that the dipolar polarization once again grows and declines according to (45) with the difference that x_0 is now to be replaced by the value of x_t at $t = t_0$.

3. References

[1] Kaplan, J.: J. Ch. Phys. *28,* 278; *29*, 462 (1958).
[2] Cobb, T. B., Johnson, C. S.: J. Chem. Phys. *52*, 6224 (1970).
[3] Apaydin, F., Clough, S.: J. Phys. C.: Solid State Phys. *1*, 932 (1968).
[4] Haupt, J.: Phys. Letters *38A*, 389 (1972).
[5] Clough, S., Hill, J. R.: Phys. Letters *49A*, 461 (1974).

Effect of Molecular Tunnelling on NMR Absorption and Relaxation in Solids

M. M. Pintar

Contents

Tunneling and NMR Absorption 128
Tunneling and Spin-Lattice Relaxation 134
References 136

About twenty years ago Tomita [1] showed that, as a result of the indistinguishability of protons, only those spin states exist in solid methane which allow the molecular wave function to be antisymmetric under the exchange of any two protons or, equivalently, symmetric under any real rotation which maps the CH_4 molecule into itself. He demonstrated that CH_4 molecules in different "spin isomer" states have very different NMR absorption spectra. The most characteristic isomeric spectrum is the delta function-like spectrum of the meta isomer (total spin two) which differs greatly from the broad and structured semiclassical absorption line. Since the spatial ground state of the CH_4 molecule combines with the meta spin isomer, it was expected that the absorption line of solid methane would remain very narrow at the lowest temperatures. This symmetry narrowing was observed in several lattices in which molecules or atomic groups are weakly hindered from reorienting [1, 2].

In addition, it has been found that tunnelling interferes with the exchange of energy between spins and lattice [3, 5]. At the lowest temperature, when coherent tunnelling of an atomic group in a solid is not interrupted by lattice phonons, its spectrum is, in the simplest case, a delta function at ω_T. In such a case, tunnelling affects in the exchange of energy between spins and lattice indirectly by partially averaging the dipolar Hamiltonian. For this reason, the spin relaxation rate at low temperatures is smaller than calculated by a semiclassical model. If coherent tunnelling is interrupted by lattice modes, its spectrum broadens with increasing temperature. Coherent tunnelling can be interrupted by "intrabarrier" transitions of the molecular vibrator. Such transition may cause spin flips if the incoherent frequency is right. This is sometimes referred to as tunnelling assisted spin-lattice relaxation.

In this overview of tunnelling effects in NMR, the four spin $\frac{1}{2}$ atomic group in various solid lattices will be considered. The three spin $\frac{1}{2}$ atomic group; e.g., CH_3, also has been studied extensively [3]. In all essential respects, tunnelling effects in CH_4 and CH_3 groups are similar.

We call the atomic rotator "semiclassical" if its degenerate ground vibrational state is split by less than the nuclear dipolar frequency ω_D (for CH_4 and NH_4 protons ω_D is $\sim 2\pi \times 4 \times 10^4$ rad sec^{-1}). In such a case different protons can be distinguished during an NMR observation of duration ω_D^{-1}, and thus the total wave function does not have to be symmetric under any real rotation which maps the rotator into itself. In other words, in such a closely-spaced tunnelling spectrum, the dipolar Hamiltonian causes strong mixing of energy states of different symmetries, rendering the concept of definite symmetry untenable.

Table I

	SEMICLASSICAL (NO SYMMETRY EFFECTS)		AND	Q.M. REGION (SYMMETRY EFFECTS DUE TO NONDISTINGUISHABILITY OF PROTONS)	
LINE SHAPE	BROAD LINE	SATELLITES & NARROWED CENTRAL LINE		SYMMETRY NARROWING: CENTRAL LINE IS NARROWER THAN PREDICTED BY THE SEMICLASSICAL THEORY	DEPOPULATION EFFECT: LINE NARROWS AS TEMPERATURE IS REDUCED
SPIN LATTICE RELAXATION	BPP FORMALISM T_1 & $T_{1\rho}$	DYNAMIC EFFECT IN THE ROTATING FRAME ($T_{1\rho}$)		DYNAMIC EFFECT IN THE LABORATORY FRAME (T_1) RATE WEAKER BY A FACTOR < 5	RATE WEAKER BY A FACTOR > 5
	NH_4Cl $E_a \approx 2000\,K$	NH_4I $(NH_4)_2SO_4$ $E_a \approx 1000\,K$	$(NH_4)_2 SnCl_6$ $E_a \approx 500\,K$		CH_4 Clathrate $E_a \approx 100\,K$
	1 2 3 ω_D		0.1 1 10 ω_L	0.1 1 10 $\frac{k}{\hbar}$	

$\omega_T \longrightarrow$

In Table I the effects of tunnelling aon absorption and on spin-lattice relaxation are grouped according to the magnitude of ω_T in a semiclassical ($\omega_T < \omega_D$) and a quantum mechanical group ($\omega_T > \omega_D$). Between these two groups is an intermediate region ($\omega_T \sim \omega_D$) in which both methods are only approximately applicable. In addition, if the tunnelling frequency is very large (extreme narrowing, ω_T is much larger than the Larmor frequency ω_L), it should be possible to observe the effect of depopulation of the higher tunnelling levels on NMR absorption.

Tunnelling effects will be briefly discussed below, beginning with the effect on the line shape, with more emphasis on the intermediate region, and continuing with spin-lattice relaxation, with an example of the extreme narrowing range.

In a consideration of the effects of tunnelling, it is helpful to know the temperature dependence of ω_T. This dependence is the result of phonon-rotor interaction. At temperatures where the reorientation (above the barrier) frequency ω_R is larger than or equal to ω_T the states of different symmetries are mixed. This is the high temperature limit in which the semiclassical approach is always appropriate. Tunnelling phenomena become important below a temperature T_T defined by $\omega_R(T_T) = \omega_T$. In many lattices T_T is less than 40 K. If hindering is moderate ($\omega_T \sim \omega_D$) then T_T is nearer 70 K. As an example, ω_T of NH_4 in NH_4I is shown in Fig. 1. In this solid the actual change of ω_T with temperature cannot be observed because, in the region where $\omega_T \approx \omega_R$, it also holds that $\omega_R \approx \omega_D$. Thus, the steep decrease of ω_T occurs in the temperature region where narrowing of the dipolar line, which is due to the random thermally activated reorientations, takes place.

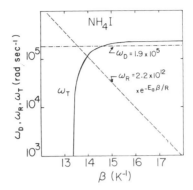

Fig. 1. Temperature dependence of ω_T and ω_R in NH_4I

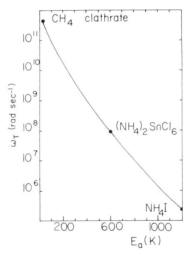

Fig. 2. ω_T vs the activation energy for reorientation for tetrahedral groups in variius lattices

It is convenient to correlate ω_T with an easily observable quantity such as the activation energy for thermally activated reorientation. In principle, however, it is not possible to get a dependable and simple functional relation between these two quantities. This is due to the sensitivity of the molecular tunnelling to the square root area of the potential barrier. For a rough estimate a plot of ω_T vs E_a should be quite useful however [6]. As an example, ω_T is plotted vs E_a for CH_4 and NH_4 in solid lattices in Fig. 2. Note that ω_T is estimated for the top two points and is probably accurate by no more than factor 5. If we use Table 1 and Fig. 2, it is possible to conclude that solids in which E_a is larger than ~ 1500 K are properly described by the semiclassical approach. In lattices where $E_a \sim 1000$ K it may be possible to observe satellites in the NMR absorption spectrum and determine ω_T accurately. If E_a is less than 1000 K, satellites are probably too weak to be detected and thus only the central, symmetry-narrowed absorption line is observed at helium temperature. For example, in ammonium iodide in the protonated sample, absorption line satellites are observed at ω_T, yet the tunnelling splitting is too small to affect the quadrupole perturbed absorption line of deuterium in ND_4I. The quadrupole splitting is $\sim 2 \times 10^5$ sec^{-1} while $\omega_T(D)$, which is roughly $1/5\ \omega_T(H)$, is of the order $2\pi \times 6 \times 10^3$ rad sec^{-1} only. Consider also the case of $(NH_4)_2SnCl_6$. Here $\omega_T(H) \approx 2\pi \times 1.5 \times 10^7$ and $\omega_T(D)$ is larger by a factor of 15 than the quadrupole splitting of $\sim 2 \times 10^5$ sec^{-1}. Consequently, the quadrupole perturbed spectrum of $(ND_4)_2SnCl_6$ is narrowed by tunnelling [7]. If E_a is ~ 100 K then the absorption line is very narrow, with a second moment of only ~ 1 G^2. In this region it should be possible to observe the depopulation of the T states. The line narrowing should occur as the temperature is reduced below ~ 1 K. Similarly the relaxation behaviour can be estimated from Fig. 2 and Table 1.

Tunnelling and NMR Absorption

If the NH_4 or CH_4 group is assumed to be rigid, its states may be classified according to the symmetry group T. The correspondence between symmetry species and total spin I is as follows: $I = 2(A), I = 1\ (T)$ and $I = 0\ (E)$. It is also assumed that the groups are in their ground electronic and vibrational states, which belong to the symmetric representation. If there is no spin-rotational interaction, the wave function can be written as a product of rotational and spin states. Since the exclusion principle requires that the total wave function must be symmetric under any real rotation which maps the group into itself, the allowed combinations are $\psi(A)\ \phi(A)$, $\psi(E)\ \phi(E)$, $\psi(T)\ \phi(T)$ which are referred to as meta, para, and ortho states, respectively. A, E, and T are labels for the irreducible representations of the tetrahedral group and ψ represents the spatical and ϕ the spin part of the wave function.

There are 16 nuclear-spin eigenstates of the Zeeman Hamiltonian represented by $|m_1 m_2 m_3 m_4\rangle$, where m_i is the z component of $\vec{I}^i$, the spin operator for proton i. m_i takes the value $\pm 1/2$ for protons. With respect to the tetrahedral symmetry group of CH_4 or NH_4 these states reduce to the members $5A + E + 3T$.

The torsional ground state of the molecule in an infinite potential well is 12-fold degenerate since there are 12 equivalent orientations of the tetrahedron. These reduce, with respect to the tetrahedral symmetry group of the molecule, to the members $A + E + 3T$. The degeneracy of the ground manifold is generally removed in a finite crystal field, but if this field has tetrahedral symmetry the three T members remain degenerate, Table 2.

For distinguishable protons there are no symmetry restrictions on the wave function, and so the nuclear-spin energy states are given by diagonalizing the complete matrix representation of the proton dipole-dipole Hamiltonian including matrix elements between all symmetry states. This is the rigid lattice "four-spin-1/2" situation and has been studied by Bersohn and Gutowski for certain crystal orientations of NH_4Cl.

Effect of Molecular Tunnelling on NMR Absorption and Relaxation in Solids   129

If the spatial energy levels are well separated; i.e., $\omega_T \gg \omega_D$, only spin transitions which conserve the symmetry type will be observed. Thus in a first-order perturbation calculation of the dipolar splittings, only that part of the matrix representation of the secular porton dipole-dipole interaction which involves matrix elements between states of the same symmetry needs to be diagonalized.

Table II

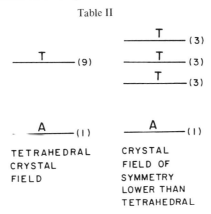

As Tomita [1] has shown for the spin isomer case, the meta component of the absorption spectrum consists of a δ function at the Larmor field H_0, while the ortho component consists of six lines symmetrically displaced from H_0, by fields which depend on spherical angular coordinates specifying the orientation of the applied field relative to the molecule.

To obtain the spectrum resulting from a powder sample a polycrystalline average must be taken of the above single-crystal spectra. The effects of neighbouring molecules can be included in an approximate way as a Gaussian broadening. The final spectrum is then the superposition of the meta and ortho components weighted according to their transition probabilities in the ratio 5 : 3 (at temperatures much larger than A to T splitting).

Proton absorption spectra of several ammonium salts were recorded [2]. Although the central components of these spectra can be reproduced quite well by a spin-isomeric line shape, the wing structure clearly cannot. Tomita's treatment predicts pronounced wing structure (ortho component) at about 9G from the center of the resonance compared with that observed at only $\sim$ 5G, and a second moment of $\sim 18G^2$ compared to several observed values of $10G^2$ or less. Some spectra show little wing structure; e.g. $(NH_4)_2 SnCl_6$, Fig. 3.

An attempt to explain these small second moments was made by assuming that the splittings of the spatial levels in such low crystal fields ($E_a \sim 500$ K) are so large that the Boltzmann factor becomes important at 4.2 K. Such a splitting would increase the population of the lower A states at the expense of the higher T ones, thereby increasing the ratio of meta: ortho (5 : 3) and hence decreasing the second moment, since the second moment of the meta component is only the result of the interionic broadening. Although this would decrease the amount of wing structure associated with the ortho component, it would still occur at the same position in the spectrum, about 9G from H_0. This situation would still be in conflict with the observed wing structure at about 5G from the resonance center in most NH_4 compounds with small E_a.

We can conclude that the spin-isomeric spectrum as calculated by Tomita is exhibited by none of the solids studied so far. Their protons, nevertheless, cannot be regarded as distinguishable. The presence of a central meta component is obvious, and its independence of crystal orientation is apparent. This is to be compared with the "four-spin-$\frac{1}{2}$" spectrum of NH_4Cl in which no such component was observed. The ortho component, which is responsible for the wing structure, is much narrower than the theoretical ortho isomer component predicted by Tomita.

One obvious mechanism for narrowing the ortho component is provided by the crystal field. As was mentioned above if the ammoniumion is situated in a tetrahedral environment, the $3T$ spatial levels are degenerate. This results in a certain line shape for the ortho component. However, if the environment of the ion has lower than tetrahedral symmetry, the degeneracy of the $3T$ levels will be removed and certain components of the ortho spectrum will be displaced by an amount equal to this splitting of the levels. If this splitting is much greater than the linewidth, these components will be pushed far out into the wings and be unobserved. This leads to a reduction of the

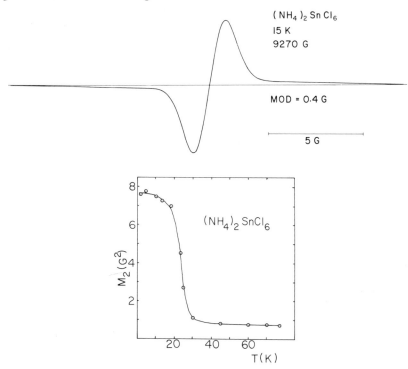

Fig. 3. Upper part: Proton NMR absorption line of $(NH_4)_2SnCl_6$ at 15 K. The narrow central peak is very pronounced (meta line). The broad wing (ortho line) is narrower than predicted by Tomita's calculation.
Lower part: The thermally activated reorientations are narrowing the dipolar line until the temperature reaches ~15 K. Symmetry narrowing is effective to 1.5 K. No depopulation effect is observed above 1.5 K

observed ortho component. On the other hand, if this splitting is only of the order of the linewidth, these components, while pushed out towards the wings, are still observed, thus broadening somewhat the ortho component. It is clear in any case that Tomita's calculation, in its failure to consider explicitly the spatial states, is inadequate to explain such an ortho spectrum.

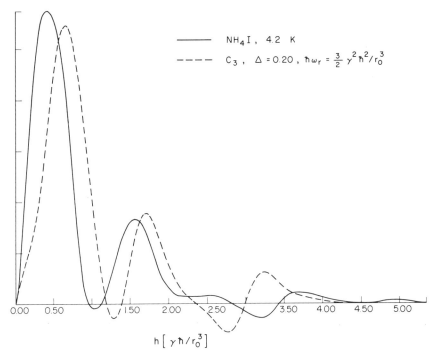

Fig. 4. The experimental and calculated (Ref. [10]) derivative of the NMR absorption in powdered NH_4I. The field h unit is the same as used by Tomita

The extreme narrowness of the meta components of most of the solids in the small E_a group should be noted, Fig. 3. An estimation can be made, for example, in the case of $(NH_4)_2 SnCl_6$ since, although the proton positions are not known accurately, the rest of the structure is. When the approximation is made that the protons are all situated at the centers of their appropriate tetrahedra, the intermolecular second-moment contribution out to third-nearest neighbours is 0.7 G^2. This is consistent with the observed broadening of the meta component of ~ 1 G^2, Fig. 3.

The most studied tetrahedral four-spin-$\frac{1}{2}$ solid is NH_4I since tunnelling satellites were observed in its NMR absorption spectrum, Fig. 4. In this lattice the molecular rotor is moderately hindered with $\omega_T \sim \omega_D$. For this reason its spectrum was reproduced with a calculation in which symmetry [8, 9] was taken into account, as well as with a semiclassical method [10].

In a semiclassical calculation, the motion of the ions is represented by a specific time dependence of the lattice variables in the dipolar Hamiltonian. In addition, the effects of the exclusion principle are ignored, a simplification which seems to be reasonable, as long as ω_r is comparable to the dipolar frequency ω_D, a condition which is certainly fulfilled in NH_4I. In other words, Tomita's truncation of the secular part of the dipole-dipole interaction does not apply in this approximation because the satellites are within the observable region of the NMR absorption spectrum. The spectrum of NH_4I was reproduced quite well by assuming that the NH_4^+ tetrahedron is rotating with a classical frequency ω_r around a specified C_3-axis, Fig. 4. It was shown [10] that this "classical" coherent [11] frequency can be identified with the tunnelling frequency of the NH_4^+ ion.

To illustrate this point let us assume that the tunneling is possible around a single C_3-axis only [10]. Under this condition the twelve-fold degenerate ground state splits into $4E + 4A$ in a manner identical to the CH_3 case [12]. The energy eigenvalues are then given by

$$E_R(A) = a + 2c ,$$

$$E_R(E) = a - c ,$$

where c represents the contribution to the rotational energy which is due to the wave function overlap. A particular member of the rotational manifold for the infinite crystal potential is denoted by $\psi_R(1, 2, 3, 4) \equiv \psi_1$, where the numbers 1 to 4 refer to the proton positions. If the NH_4-tetrahedron is rotated by $\pm 2\pi/3$ around the C_3-axis passing through the proton 4, two other wavefunctions belonging to the 12-fold degenerate set are: $\psi_R(2, 3, 1, 4) \equiv \psi_2$ and $\psi_R(3, 1, 2, 4) \equiv \psi_3$. For the purpose of this discussion, only these three states will be considered.

If we allow for the tunnelling around a chosen C_3-axis, the set ψ_1, ψ_2, ψ_3 will split into one A state and two E states. The corresponding eigenfunctions are [12]

$$\psi_R(A) = N_R(\psi_1 + \psi_2 + \psi_3) ,$$

$$\psi_R(E_1) = N_R(\psi_1 + \epsilon\psi_2 + \epsilon^*\psi_3) ,$$

$$\psi_R(E_2) = N_R(\psi_1 + \epsilon^*\psi_2 + \epsilon\psi_3) ,$$

and the energy eigenvalues are given above. N_R is the appropriate normalization factor and $\epsilon = e^{i2\pi/3}$. An arbitrary state within this three-dimensional subspace describing the NH_4^+ ion tunnelling around this single C-axis can be written, neglecting normalization, as

$$\psi(t) = C_1 e^{-\frac{i}{\hbar}E_R(A)\cdot t} \psi_R(A) + C_2 e^{-\frac{i}{\hbar}E_R(E)\cdot t} \psi_R(E_1)$$

$$+ C_3 e^{-\frac{i}{\hbar}E_R(E)\cdot t} \psi_R(E_2) ,$$

where the C_i's are complex numbers. Introducing $\hbar\omega_T \equiv E_R(E) - E_R(A)$, $\psi(t)$ becomes in terms of ψ_1, ψ_2 and ψ_3:

$$\psi(t) = e^{-\frac{i}{\hbar}[E_R(A)+E_R(E)]\cdot t/2} \left\{ \left[C_1 e^{+i\omega_T t/2} + (C_2 + C_3) e^{-i\omega_T t/2} \right] \psi_1 \right.$$

$$+ \left[C_1 e^{+i\omega_T t/2} + (\epsilon C_2 + \epsilon^* C_3) e^{-i\omega_T t/2} \right] \psi_2$$

$$\left. + \left[C_1 e^{+i\omega_T t/2} + (\epsilon^* C_2 + \epsilon C_3) e^{-i\omega_T t/2} \right] \psi_3 \right\} .$$

If we denote the initial condition

$$\psi(t = 0) = \psi_s, \text{ where } s = 1, 2, 3,$$

then the above $\psi(t)$ simplifies to

$$\psi(t)_s = e^{-\frac{i}{2\hbar}[E_R(A)+E_R(E)]\cdot t} \sum_{r=1}^{3} \left[\frac{1}{3} e^{+i\omega_T t/2} + \left(\delta_{r,s} - \frac{1}{3}\right) e^{-i\omega_T t/2}\right] \psi_r$$

where s characterizes the initial condition. If the measurement of the orientation of the NH_4^+ ion is performed at the time τ, then the probability $p_{r,s}(\tau)$ that the NH_4^+ is found in the state with definite orientation ψ_r is $p_{r,s}(\tau) = |\langle\psi_r|\psi(\tau)_s\rangle|^2$. If we assume that there is a small overlap; i.e., $|\langle\psi_r|\psi'_{r'}\rangle| \ll 1$ where $r \neq r'$, the following probability is obtained:

$$p_{r,s}(\tau) = \delta_{r,s}\left(\frac{1}{3} + \frac{2}{3}\cos\omega_T\cdot\tau\right) + \frac{2}{9}(1 - \cos\omega_T\cdot\tau).$$

The corresponding classical correlation function $G(\tau) = F_{ij}^{(0)} F_{ij}^*(\tau)$, where

$$F_{ij}(t) = -\left[\frac{8\pi}{5}\right] \sum_{m'\neq 0} Y_2^{m'}(\theta'_{ij}, 0) Y_2^{m'*}(\beta, \gamma) e^{im'\phi'_{ij}(t)},$$

is proportional to

$$G(\tau) \propto \sum_{r,s} p_{r,s}(\tau) e^{im'\phi_r} e^{im''\phi_s}.$$

In this expression $m', m'' = \pm 1, \pm 2$ and $\phi_s = 0, \pm 2\pi/3$ or $3\pi/2, 3\pi/2 \pm 2\pi/3$. In any case, $G(\tau) \propto A + B\cos\omega_T\cdot\tau$, where A and B are constants. If we neglect the constant term A, the Fourier transform of $G(\tau)$ is $J(\omega) \propto \delta(\omega + \omega_T) + \delta(\omega - \omega_T)$, where δ is the Dirac delta function. With this spectrum, the transition probability per unit time becomes equal to the semiclassical transition probability with $\omega_R = \omega_T$. Therefore, the frequency of rotation, ω_r, used in the semiclassical calculation of the absorption spectrum, is to be interpreted as the tunnelling frequency of the NH_4^+ ion. It should be noted that classical rotation also generates satellites at $\omega_0 \pm 2\omega_r$. This is due to the terms $e^{\pm i2\phi}$, where ϕ varies continously as $\phi = \omega_r t$, which appear in the expression for $F_{ij}(t)$.

On the basis of the preceding discussion, it is not surprising that the lineshape of NH_4I can be reasonably approximated by assuming that the NH_4^+ ion rotates with a specific classical frequency about a symmetry axis, since this frequency is a measure of the tunnelling frequency ω_T.

If ω_T is so large that only the central, symmetry narrowed part of the absorption spectrum can be detected, it does not seem possible to deduce ω_T directly by analyzing such a spectrum at any one temperature. It would be possible to obtain this information from the temperature dependence of the absorption line if the depopulation of the first excited state could be observed. However, this depopulation effect can be seen above 1.3 K only if the first excited state is $\gtrsim 1$ K above the ground state. It is apparent from Fig. 2 that only those lattices in which $E_a \sim 100$ K may show depopulation.

Tunnelling and Spin-Lattice Relaxation

The effect of tunnelling on spin-lattice relaxation is more complicated than its effect on line shape [3, 5]. If the hindrance is moderate as in NH_4I, a $T_{1\rho}$ minimum near $\gamma H_1 \sim \omega_T$ is expected in the rotating frame. This leads to a very complicated $T_{1\rho}$ dispersion which has its origin in mixing of the Zeeman and dipolar energies at $H'_L \leqslant H_1 \leqslant 3H'_L$, in cross relaxation between iodine and proton spins as well as in the supposedly Debye spectrum of random reorientations. Since $H'_L \sim 4G$, it takes an H_1 of 20G to decouple the Zeeman and dipolar energies. This leaves only a small range for a dispersion study without mixing problems. Theoretical and experimental work is currently being undertaken by R. Hallsworth in this direction.

In lattices where hindrance is smaller and $\omega_T \gtrsim \omega_L$, in many instances two minima are observed if the solid is cooled from ~ 100 to ~ 10 K, see for example Fig. 5. This behaviour has been studied thoroughly [4, 3].

If hindrance is even weaker (extreme narrowing range) only one T_1 minimum is observed. A good example is a CH_4 molecule trapped in D_2O clathrate.

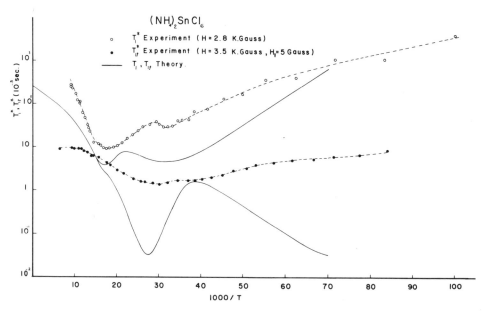

Fig. 5. Proton spin-lattice relaxation of polycrystalline $(NH_4)_2SnCl_6$ at high and low fields. The experiment is compared with the results of a semiclassical calculation in which both thermally activated reorientations around both C_2 and C_3 axes were assumed to relax nuclear spins. Note that the magnetization decay is nonexponential and the relaxation times presented are 1/e values. (J. Peternelj, Ph. D. thesis, 1973, unpublished)

A clathrate is one of a class of solids in which small molecules, such as CH_4, occupy almost spherical holes in an ice-like lattice of hydrogen bonded water molecules. The sample which was studied [3] is $2CH_4 \cdot THF\text{-}d8 \cdot 17D_2O$, a double deuterohydrate of CH_4 and deuterated tetrahydrofuran. The sample was maintained at each temperature for one hour prior to a measurement, and the T_1 measurements were taken in order of decreasing temperature. This type of experiment is referred to as a "slow cooling run", with the time taken to cool from 80 K to the helium boiling point being of the order of 6–8 hrs.

At 15 MHz T_1 has a minimum of 195 msec at 5.4 K in a slow cooling experiment. The apparent activation energy on the high temperature side of the minimum is 24 K, while on the low temperature side it is 7 K. Assuming that random isotropic reorientation is the relaxation mechanism near the T_1 minimum, and using the high temperature slope, standard semiclassical analysis gives $8.5 \times 10^{-11} \exp(23.5/T)$ sec as the correlation time for the motion. On the high temperature side of the minimum, T_1 is frequency independent and varies exponentially with temperature. Both of these features are in agreement with the semiclassical theory for relaxation through modulation of the intramolecular dipole-dipole interaction by random isotropic reorientation. However, the semiclassical model, when applied to CH_4 molecules, predicts a T_1 minimum of 4 msec at 15 MHz. This is less than the observed value for the CH_4 clathrate by a factor of 50.

In a separate experiment, the sample was cooled quickly (15 minutes) from 80 K to 5.4 K. The measured T_1 was only 145 msec, 50 msec less than in the slow cooling run, and did not change over a period of 5 hrs. When the sample was subsequently slowly warmed, T_1 approached the slow cooling value, becoming equal to it at 7.5 K. This behaviour suggests that below 7.5 K there are no transitions between the A and T states, so that their populations are "frozen" in a non-equilibrium ratio. If this is the case, the effective temperature of the rotational levels at the T_1 minimum would be about 7.5 K for the slow cooling experiment.

Since the matrix elements of the intramolecular dipolar Hamiltonian vanish within the manifold of $I = 2$ spin states, a direct spin-lattice coupling occurs only within the $I = 1$ (T) states. It has been suggested that the A spins relax via the T spins [5]. The observation of a single spin-lattice relaxation time in the CH_4 clathrate seems to support this assumption. Accordingly, the relaxation rate of the entire spin system is

$$T_1 = T_1(T+A) = \left\{ 1 + \frac{5}{3} \exp[(E_T - E_A)/kT] \right\} T_1(T).$$

$T_1(T)$ has been calculated for CH_4 by A. J. Nijman, using free rotor $J = 1$ wave functions. The minimum value at 15 MHz was found to be 125 msec. It is not possible to adjust the parameters in the above equation to obtain agreement of his calculated value with the experimental T_1. This indicates that, even though the hindrance to rotation is very small, the free rotor wave functions are not adequate to describe the CH_4 molecules in this lattice.

Comparative analysis of the fast and slow cooling experiments may give some idea of the actual minimum of $T_1(T)$, as well as of the splitting between the T and A rotational levels. Let us assume that in the fast cooling experiment the effective temperature of the rotational states is very high, so that the exponential in the above equation is unity, and that the effective temperature in the slow cooling experiment is 7.5 K. This gives a $T_1(T)$ minimum of 54 msec and an A to T splitting of 3.4 K. According to the above analysis, these values should be considered as upper and lower limiting values of the $T_1(T)$ minimum and of the splitting respectively.

References

[1] Tomita, K.: Phys. Rev. *89*, 429 (1953).
[2] a) Sharp, A. R., Vrascaj, S., Pintar, M. M.: Solid State Commun. *8*, 1317 (1970).
[2] b) Watton, A., Sharp, A. R., Petch, H. E., Pintar, M. M.: Phys. Rev. *B5*, 4281 (1972).
[3] Clough, S.: Lecture presented in this volume and references therin.
[4] Haupt, J.: Lecture presented at the Third Waterloo NMR Summer School (1973). Also Z. Naturforsch., *26a*, 1578 (1971).
[5] deWit, G. A., Bloom, M.: Can. J. Phys. *47*, 1195 (1969).
[6] Smith, D.: J. Chem. Phys. *58*, 3833 (1973).
[7] Knispel, R. R., Petch, H. E., Pintar, M. M.: Solid State Commun. *11*, 679 (1972).
[8] Watton, A., Petch, H. E.: Phys. Rev. *B7*, 12 (1973).
[9] Dunn, M. B., Ikeda, R., McDowell, C. A.: Chem. Phys. Letters *16*, 266 (1972).
[10] Peternelj, J., Hallsworth, R., Pintar, M. M.: Phys. Rev. *B1*, Aug. (1976).
[11] Andrew, E. R.: Progress in Nuclear Magnetic Resonance Spectroscopy *8*, 1 (1971), see also the references therein.
[12] Clough, S.: J. Phys. C.: Solid State Phys. *4*, 2180 (1971).
[13] Nicoll, D. W., Peternelj, J., Davidson, D. W., Hallsworth, R. S., Pintar, M. M.: Phys. Letters *55A*, 127 (1975).

How to Build a Fourier Transform NMR Spectrometer for Biochemical Applications

A. G. Redfield

Contents

I. Introduction 137
II. What is FTNMR? 138
III. Correlation Spectroscopy 139
IV. Problems in FTNMR 139
V. Solutions 141
VI. Artifacts, Limitations, and Comparisons 148
VII. Acknowledgements 150
Appendix A. Aspects of Conventional FTNMR 150
Appendix B. Phase and Amplitude Corrections in FTNMR 151
References 152

I. Introduction

The low sensitivity of NMR, relative to other analytical methods, may prevent it in the long run from becoming the central tool in biochemistry that it is in chemistry. However, NMR is becoming more and more used in this field. This trend can be confirmed by glancing at any current issue of a biochemistry journal. In the present chapter the development of the basic tool of this work, the high resolution NMR spectrometer, is described. Emphasis is on proton resonance because the high sensitivity for protons relative to other nuclei makes them attractive despite other limitations. I will not discuss magnets; with a few minor exceptions, it is desirable to have as large a magnetic field as possible.

When I started to work in this area there was essentially one published description of a spectrometer by Ernst and Anderson [1, 2] who originated the idea of FT (Fourier transform) NMR as a significant and practical improvement over frequency-swept NMR. I felt that their design could be improved considerably, and proceeded to build my own modified version, with the help of Raj K. Gupta and others [3]. I had two motivations: I wanted to use the instrument for research in biochemistry, and I wanted to constructively influence the design of commercial spectrometers which were then being developed.

We were successful in the first goal, but failed completely in the second: even though we published a thorough description of our spectrometer in a prominent serial [3], gave a paper at the annual experimental NMR meeting on it, and published several unique research articles, commercial spectrometers are only now beginning to depart significantly from Ernst and Anderson's original scheme. This failure may be due in part to the fact that our detailed account did not include a separate detailed description of those features of the instrument which were fundamentally novel, as opposed to those which were engineering details. The present article is for the most part an attempt to do this. You should consult our original articles for details [3].

II. What is FTNMR?

A strong (~ 10 μsec) pulse flips all the spins by 45 to 90°, and their precessing magnetic moment induces a decaying radio frequency signal in a sample coil. A simplified block diagram of a "conventional" FTNMR spectrometer is shown in Fig. 1a. For simplicity I assume an operating frequency of 100 MHz throughout this chapter, and omit the field stabilization system based on deuteron or fluorine NMR. The NMR rf frequency at ~ 100 MHz is too high to be processed directly by a computer, so it is mixed with the transmitter frequency to produce an audio beat note, shown in Fig. 1b. This

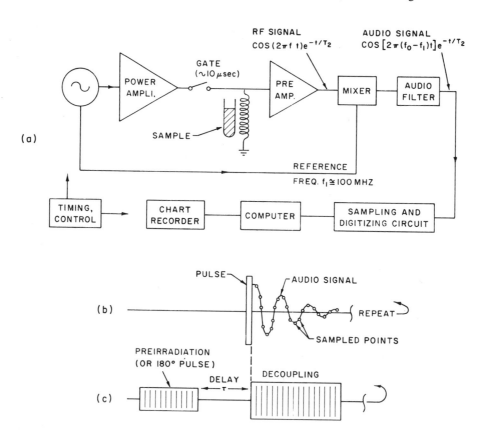

Fig. 1. (a) Block diagram of a conventional FTNMR instrument, after Ernst and Anderson. (b) Observation pulse, audio signal, and signal sampling. (c) Double irradiation sequence. If the preparatory pulse were a 180° pulse it would, of course, be shorter than suggested here. Decoupling is time-shared, if homonuclear.

signal is filtered to reduce higher frequency noise, sampled at a high audio frequency, digitized, and added to previously accumulated signals to increase the signal-to-noise ratio. The accumulated signal is Fourier transformed, and the complex Fourier transform is then phase-rotated in the complex plane by the computer, under operator control, to compensate for phase shifts produced by the rf circuitry and audio filter. Finally, the real part of the transform is plotted, and has the same shape as the swept NMR spectrum. The advantage over frequency swept NMR is that all the nuclei emit a signal at once, whereas in the swept mode much time is necessarily wasted searching between narrow resonance lines.

Double irradiation experiments, symbolized in Fig. 1c, are an important feature of any good spectrometer. Preirradiation is used for measurements of T_1, transfer of saturation, and other sophisticated experiments. Decoupling is usually noise modulated if heteronuclear, or is time shared if homonuclear, with time sharing in synchrony with the data sampling. Further minor improvements in this basic design are discussed in Appendix A.

III. Correlation Spectroscopy

In competition with FTNMR is a roughly equally sensitive class of techniques known as correlation spectroscopy. The single-pulse excitation is replaced by a carefully designed stimulation lasting a longer time. Two types have been used: a computer-designed pseudo-random sequence of relatively weak pulses [4]; or a rapidly frequency modulated pulse which scans the spectral region of interest (called rapid scan NMR) [5]. The signal from the nuclei is cross-correlated with the stimulus and this correlation function is Fourier transformed. After a phase shift to compensate for filtering, the real part of the transform is plotted. If the NMR response is linear (i.e. proportional to the stimulus) then linear response theory shows that this plot is identical to the swept NMR absorption spectrum.

The rapid-scan version of correlation NMR has been used in several laboratories. It is attractive because it can be adapted to a swept spectrometer with relatively little hardware modification other than addition of a digitizer and computer. It is also possible to sweep only a part of the spectrum, avoiding strong solvent lines, for greater dynamic range.

IV. Problems in FTNMR

a. Imaging and Aliasing. In a spectrometer constructed as in Fig. 1a, the transmitter frequency must always be "placed" at one side of the spectrum (Fig. 2a); i.e. it should be lower than the frequency of the lowest field line, or higher than the highest. That is because a line 100 Hz higher than the carrier frequency f_1 gives the same audio signal as one 100 Hz lower, namely, a 100 Hz audio decay. Thus, a line at a field point a in Fig. 2a is imaged about f_1 to a'. This is *imaging*.

Furthermore, a frequency which is higher in audio frequency than half the sampling rate ν_a will appear to be a lower frequency after sampling (Fig. 2b). Without going into a general proof, we state that any line on the frequency scale is *aliased* about mirror image points $\pm 1/2\nu_a$ from the carrier frequency. Thus, a line at b is aliased to b'. Aliasing is avoided by choosing a sampling rate greater than twice the width of the entire spectrum.

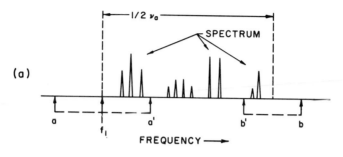

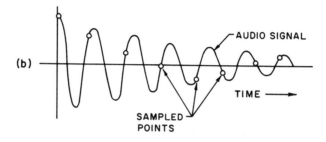

Fig. 2. (a) Conventional placement of the transmitter frequency relative to a spectrum. The points a, a', b, b' are discussed in the text. The frequency ν_a is the repetition rate at which the signal is sampled for digitization and accumulation by the computer. (b) The result of sampling a high frequency signal at a low sampling rate. The dots represent sample points.

This strategy works as long as you know where the spectrum is; in practice it sometimes leads to confusion when a change of solvent lock-compound shifts the entire spectrum across either f_1 or $f_1 \pm (1/2)\nu_a$. More serious, the distance between output points on the chart is the advance frequency ν_a divided by the number of input points. For protons at 270 MHz, $(1/2)\nu_a \approx 3$ kHz as shown in Fig. 2a, so if the computer can only handle 8096 input points, as is common, then the maximum resolution is about 1 Hz. The problem is worse for carbon, fluorine, and phosphorous. The computer-limited resolution can be increased by a factor of about two (or in some cases slightly more) by setting f_1 close to one end of the spectrum as in Fig. 2a, and choosing $1/2\,\nu_a$ equal to about one half to two thirds of the total width of the spectrum. If the filter is set to cut off very sharply at an audio frequency slightly less than $1/2\,\nu_a$, then aliasing will not be serious, provided there are no very strong lines in the half of the spectrum that is supposedly not being observed. The spectrum is then obtained in two halves — a high frequency and a low frequency half — one of which will be plotted backwards unless the computer program is altered.

Obviously, the best solution to this problem is to get a bigger computer. However, the core size needed for fluorine or phosphorus is forbiddingly large.

b. Sensitivity. A more serious problem is that noise, as well as signal, can be imaged and aliased from outside a spectral region, into it. Noise imaging gives a factor of two in noise power over that present without imaging, with a corresponding two-fold increase in running time needed for a given signal to noise ratio. Noise aliasing is reduced by the filter in Fig. 1a which is usually designed to cut off rapidly at an audio frequency just above $1/2\ \nu_a$. On commercial instruments this cutoff is sometimes designed to be unnecessarily gradual, to avoid the strange-looking variation of noise near the right-hand edge of the chart, at $1/2\ \nu_a$, which occurs for a well chosen filter. Thus, aliasing also increases noise power by a variable amount up to about twofold, depending on the filter design and on the line position, with a corresponding multiplication in running time. The combined effect of imaging and aliasing of noise can thus be up to four-fold inefficiency.

c. Dynamic Range. It is often desirable to do biochemical NMR in predominantly H_2O (rather than D_2O) solvent — either because it is inconvenient to change the solvent or because one whises to observe protons which are slowly exchanging with solvent (amide protons, for example). These protons are present in sub-millimolar concentrations in some cases, whereas water protons are at a concentration of 110 molar. Thus one must have a system with high dynamic range. The required dynamic range is not obtained by calculating the solute-to-solvent ratio as just implied, but rather by calculating the solvent signal-to-noise ratio going into digitizer, for typical filter bandwidths. For H_2O this is several thousand at 100 MHz, and higher at 220–360 MHz. It is conceivable that digitizers of the required resolution and speed will be available, but there would also be problems associated with the rest of the data-handling process.

One useful solution to this problem is to pre-saturate the solvent signal and not those of the solute (see below). In some biochemical experiments this method is not acceptable: if the exchange rate of a solute proton equals, or exceeds, its relaxation rate $(T_1)^{-1}$, then its resonance intensity will be decreased, or its resonance eliminated from the spectrum, when the water line is saturated.

V. Solutions

a. Quadrature Detection and Optimal Filtering. The principles of single-sideband generation and detection, as a method of noise elimination and efficient signal handling, have been known to radio amateurs for years. They provide straightforward methods for eliminating the imaging problem. One method, crystal filtering, will be discussed briefly later. The other method, quadrature detection, is the one which we used and which I now describe. I would like to stress that our important accomplishment was not simply the implementation of this fairly obvious technique in NMR. Rather it was the demonstration that this could be done straightforwardly with low imaging problems, and most importantly in conjunction with versatile audio filtering and a method for phase and amplitude compensation to almost completely eliminate aliasing with only a slight loss in resolution. In addition to providing the expected increase in productivity because noise is no longer aliased and imaged, this technique permits instruments with small core memory to take very high resolution sections of parts of a spectrum; it is an almost-essential requirement for implementation of long pulse methods; and as was mentioned above it eliminates the frequency-proportional phase shift which must be manually controlled on current instruments. Optional filtering requires no extra hardware when designed properly; phase compensation requires about 300 computer words and adds about 10 % to the time required to do an FT. Phase compensation is not absolutely necessary, but is very desirable.

These improvements can be added on to existing instruments [6–9] and are now becoming commercially available (at least quadrature detection is). However, they should always be an integrated part of the instrument, with the filter automatically switched to match the sampling rate, unknown to the operator.

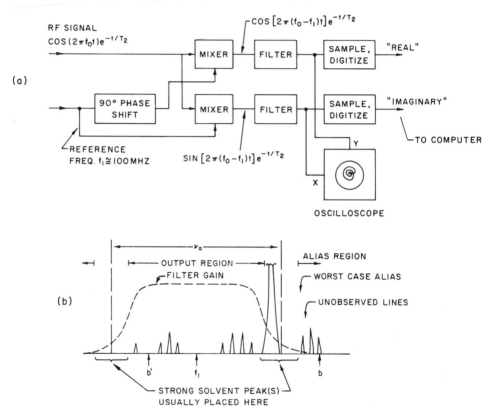

Fig. 3. (a) Quadrature phase detector. In practice a single digitizer is used in conjunction with two sample and hold circuits and an analog input multiplex. (b) Position of the usable part of the spectrum in quadrature detection. Also shown is the usual position of the solvent line for the long pulse techniques

Quadrature phase detection is shown schematized in Fig. 3a. A single phase detector, detecting a single free induction decay, puts out a signal $\cos[2\pi(f_0 - f_1)t + \phi]$, where ϕ is some arbitrary phase which depends on the instrument phase shifts. As we mentioned above, the computer cannot tell whether originally $f_0 > f_1$, or $f_0 < f_1$. A second phase detector, in quadrature with the first, puts out an audio signal 90° out of phase, $\cos[2\pi(f_0 - f_1)t + \phi - \pi/2]$ which is the same as $\sin[2\pi(f_0 - f)t + \phi]$. The two outputs of the quadrature phase detectors can be thought of as real and imaginary parts of a complex signal $\exp[i2\pi(f_0 - f_1)t + \phi_2]$; and they are treated as such by the computer. Obviously the computer can now distinguish the case $f_0 > f_1$ from $f_0 < f_1$; if this is not clear to you, consider what the oscilloscope pattern (Fig. 3a) looks like in these two cases and see Ref. [*3*].

If the two phase detectors, filters, and digitizers are identical, and the phase shift exactly equal, then there will be zero imaging. If there is a 1 % gain difference, or a .01 radian (1/2 degree) phase shift difference, between the two channels (or in the nominal 90° phase shift) then there will be an image of size only 1 % that of its source. At this writing at least four quadrature detectors and filters have been constructed which achieve at least this level of perfection. In the two constructed in my labs, by myself and Mrs. Sara Kunz, the signal was first converted to 50 or 100 kHz before quadrature detection which is based on high speed FET switches. All relevant circuits are made with operational amplifiers, 1 % resistors, and manually matched capacitors, and the phase shift finally trimmed manually, yielding a system which operates without adjustment for years.

Even better balance is achieved (and/or less excellent components are needed) if a scheme called "data routing" is used [*7*]. After every two transients, the transmitter phase, or the incoming reference phase, is shifted 90°, thus shifting the complex signal into the computer by 90°. The computer is programmed to shift the signal's phase mathematically before adding it to memory, in the opposite direction. This effectively interchanges the real and imaginary channels, averaging their properties to first order. Thus, if the two channels are already matched to 1 % and .01 radian, image levels of much less than 10^{-3} should be possible.

The real and imaginary signals can be multiplexed and sampled sequentially rather than simultaneously, and we have described two schemes for doing this [*3, 8*]. These schemes are compatible with data routing and optimal filtering. I will not discuss them here because it is usually faster to sample and digitize simultaneously. One of these schemes is most useful for simple conversion of some existing instruments to dual detection [*8*].

Since imaging is not a problem, the rf frequency f_1 is set in the center of the spectrum as in Fig. 3b. Aliasing can still occur, and produces a spurious signal at b' if there is a line at b, shifted by $\pm \nu_a$ (which is the sampling rate per *pair*, real and imaginary, of input signals. If real and imaginary signals were sampled sequentially, then ν_a is *half* the rate of sequential sampling). The transmitter power need now be only 1/4 what it would be if f_1 were at the side of the spectrum, a minor advantage.

Aliasing is almost eliminated in our present spectrometer by using sharp cutoff 6-pole (Butterworth) filters which have half gain points at a frequency of about 4/5 of that of the alias limit, $1/2 \nu_a$. A band of which 4/5 of the maximum non-aliasing width (ν_a) is thus passed to the computer. The computer-limited resolution is thereby decreased only by about 20 %. After phase and amplitude correction this section of the spectrum is plotted, while the remaining 1/5 is discarded. The worst case (for aliasing) comes from lines slightly more than $\pm .6 \nu_a$ from f_1. These lines are passed by the filter with a gain of about 1/30 that of the maximum frequency lines just inside the range $\pm .4 \nu_a$, so the maximum alias level is 3 %. Even smaller levels are achieved using long pulse methods (next section). In practice, only strong solvent lines give troublesome aliases.

Turning to the problem of phase and amplitude correction of the filter distortion, it is not absolutely necessary to do this automatically. Manually controlled frequency-proportional phase rotation can be used, as is usual on current commercial instruments. However, amplitude distortion will still occur at the edges of the spectrum, and as we already indicated there is no good reason not to make a precise completely automatic correction.

Automatic correction could be implemented by measuring the gain G and the phase shift ϕ of one of the filters, as a function of frequency, then storing in the computer a complex function $G^{-1}e^{-i\phi}$, and multiplying the output (complex) FT by this function. In practice, it is easier to measure the transient response of the filter and Fourier transform it to get G and ϕ. It is also desirable to store the result in the computer memory in sampled form for later interpolation when the phase correction is made. The method for doing this is described elsewhere [3] and in more detail in Appendix B. In our present spectrometer the filter bandwidth is changed ten-fold by relay switching of capacitors, and in steps of 1, 2, 4 and 8 by relay switching of resistors. The sampling frequency ν_a is switchable between 128, 256, 512, and 1024 Hz, and ten times these, and the output spectral width is one or ten times 100, 200, 400, or 800 Hz. The filter bandwidth is relay switched to scale proportionally with the sampling rate, so that the same computer correction is always used. A greater maximum spectral width and a finer mesh of choices within this width would be feasible and desirable.

Provided there are no large signals near 0.6 ν_a to give aliasing, *the filter correction described is perfect* in the sense that the plotted spectrum is the real part of the FT of the incoming transient, over the selected spectral range. Even broad lines, whose tails extend outside the plotted range, are faithfully represented.

In conclusion, we discuss briefly the alternative to quadrature detection, namely crystal filtering of the rf signal [9]. This method is clearly simpler when a single spectral width is used and precise intensity measurement is unnecessary. It seems less attractive for a multipurpose spectrometer because of the expense of having a variety of filters for different bandwidths. However, the method of Appendix B could be used to measure each filter characteristic, and by filtering at an intermediate frequency whose center band is constant, the observed spectral window could be shifted about. To my knowledge, no one has built a multipurpose instrument based on crystal filtering, so the relative merits are hard to assess.

b. Long Pulses. The previous section was devoted to methods for *detecting* a signal from some specific frequency range without spurious sensitivity to other regions. The present section concerns methods for selective *stimulation* of a spectral region and smaller or zero stimulation of other regions or places on a spectrum. We sought a way to do this while losing as little as possible of the time-resolvent stimulation which is such an attractive feature of FTNMR.

The simplest way of doing this was pointed out long ago by S. Alexander [10], and exploited since then in certain non-FT pulse experiments [11]. This is to use a relatively long weak pulse whose frequency is in the center of some region of interest, and is some distance ν_s away from the solvent resonance — usually 100 Hz or more. The pulse length

How to Build a Fourier Transform NMR Spectrometer for Biochemical Applications 145

τ is chosen to be slightly less than ν_s^{-1}, and its amplitude then chosen to make it a 45° to 90° pulse for spins in exact resonance with it.

Insofar as the spins respond linearly (which would be true if the pulse were much weaker than 45°), the pulse effectiveness would be roughly proportional to the magnitude of its FT, which is a constant times $(\sin x)/x$, where $x = 2\pi(f_0 - f_1)\tau$ (Fig. 4a). This has nulls at frequencies given by $|f_0 - f_1| = \tau^{-1}$. Thus, if the pulse length τ is equal to ν_s^{-1} there should be a null in its effectiveness at the solvent frequency, in the linear approximation.

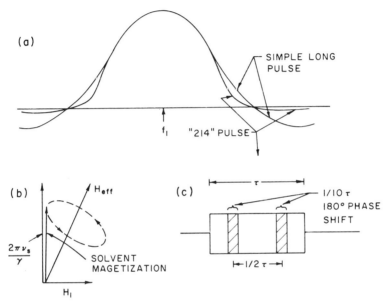

Fig. 4. (a) Fourier transforms of a simple long pulse, and of the 214 pulse, normalized to the same value at center band. (b) Solvent effective field for the long pulse. The initial (and final) position of solvent is shown, and its trajectory during the pulse is indicated by the dotted line. (c) The 214 pulse. Its amplitude is about 65% greater than that of the simple long pulse, in order to produce the same flip angle at center band f_1.

This is very nearly correct for a finite pulse amplitude. To be more quantitative we must make use of the effective field in the rotating frame. The effective field is just H_1 for spins at the center band, and $\gamma H_1 \tau = \pi/4$ for a 45° pulse. For the solvent the effective field (Fig. 4b) is the vector sum of H_1 and the amount by which the water is off resonance, in field units. Its magnitude H_{eff} is $[H_1^2 + (2\pi\nu_s/\gamma)^2]^{1/2}$. To null the solvent, the pulse length must be such that solvent is nutated back to its starting point. The condition for this is $\gamma H_{\text{eff}} \tau = 2\pi$. For a 45° pulse at resonance these equations require that ν_s be $(63/64)^{1/2}$, or about 0.99, of τ^{-1}. Spins intermediate between center band and solvent get flipped an amount intermediate between 45° and zero, and their orientation is not perpendicular to H_1 in the rotating frame. Thus the phase of lines due to these spins will be incorrect, and a phase and amplitude correction must be made. This is straightforward and can be combined with the filter correction routine described in the previous section. In our present instrument we have a library of up to nine such corrections corresponding to different ratios of ν_s to the sampling frequency ν_a, and all for 45° pulses. In practice the three ratios $\nu_s = 0.8$, 1.0, and 1.2 times $1/2\,\nu_a$ are almost always used because then the solvent is never imaged onto the plotted spectrum (see Fig. 2).

Once the system is set up with a high quality rf attenuator, digital pulse timer, and a phase and amplitude correcting computer program, it is easy to use, even for inexperienced operators.

Whereas a short-pulse spectrum after filter correction is perfect, with constant noise across the band, the long pulse spectra show extra noise toward the edges of the spectra because the signal is smaller in these regions, and when this is compensated for by the computer routine, the noise is increased. Very broad lines are not properly reproduced, also in contrast to short pulse spectra; in practice this is the source of a baseline curvature observed when there is a strong (protein) line which is at the edge of the plotted region. However, for ordinary spectra consisting of only moderately broad lines (e.g., amide resonances up to 50 Hz wide) the spectra are nearly perfect.

Preirradiation is an important part of FTNMR's capability, and presents problems in protonated solvents. After even a weak preirradiation there will be a transient from the solvent, which gives an annoying random baseline curvature even if it does not overload the digitizer. This can be reduced to an acceptable level by a homogeneity spoil pulse. We also find it useful to gate the preirradiation off slowly (in ≈ 5 ms), which reduces the post-transient by returning the preirradiation effective field to the axis adiabatically. We always use the saturation-recovery technique for T_1 measurements; use of a 180–90°, or 90-spoil-90° sequence would be difficult, though possible, but saturation-recovery seems easier.

In macromolecules we often look at lines at the edge of the spectrum, and place f_1 so that the central aliphatic and aromatic parts of the spectrum are in or near the unobserved and unaliased region between 0.2 and 0.3 ν_a (Fig. 2b) from f_1, with the solvent H_2O peak nulled out with a long pulse. There is generally a sizable baseline curvature in such spectra which results not from the 110 M H_2O protons, but from the one to five total moles per liter of aliphatic and aromatic protons. These are weakly stimulated by the long pulse, and have a dispersive wing extending into the region around f_1. Just after the pulse these protons are oriented at an angle (in the rotating frame) other than perpendicular with respect to the rotating rf field, and, in the output spectrum, their dispersion tail is misphased to give an apparent contribution to the absorption at center band.

To make this problem less severe, we looked for a pulse which has a null effect in a region of the spectrum, rather than a point, as far as possible. We required that our pulse be easy to generate, preferably using only discrete phase modulation, rather than difficult-to-achieve amplitude modulation. Fig. 4c shows a very useful pulse of this kind which is an approximation to a plain pulse of length τ added to two very short phase reversed pulses [12]. The strengths of these are chosen to give a nearly zero derivative, as well as a zero value of the pulses FT, at a distance τ^{-1} from the pulse frequency. We assumed that a low FT value in this region will result also in low NMR stimulation over the same region. We call this a "214" pulse, after its first three intervals, and show its FT in Fig. 4a. Generally the intervals are set up as shown, with the amplitude such that spins at center-band are still flipped by about 45°, and the solvent frequency is set a distance about τ^{-1} from f_1. Then the pulse intervals are fine-tuned empirically to give the flattest baseline; this is usually about five times as flat as that attainable with the simple long pulse. The same phase/amplitude correction is used as for the simple long pulse, though this may give some amplitude distortion at the spectrum edges. In practice the method is much easier to use than it sounds.

Our previous publications have apparently given the impression that our methods permit observation of only narrow regions of the spectrum, far from solvent. However, we can routinely work up to 50 Hz from solvent with solvent saturation, and we routinely observe lines 100 Hz from water without saturation, at 90 MHz, though we do have to shim carefully and be on the alert for spinning sidebands. The width of spectrum observable with a single spectrometer setting is about four times the distance of the closest observed line to the solvent (see Fig. 3b). We can automatically scan symmetrically on both sides of solvent water, in successive runs, and only occasionally do we need two runs of different frequency and spectral width to cover the entire downfield region.

We have also used these methods with protonated organic solvents such as ethanol which present difficulties because there is no single solvent line. We have been able to study aromatic protons in aliphatic solvents by combining all the techniques outlined, including solvent saturation with a frequency modulated prepulse (see below). In general these cases are difficult, require considerable experimentation, and one cannot be assured of success using long pulse methods.

c. Water Elimination FT ("WEFT") and related methods eliminate the solvent signal by destroying the solvent magnetization, as far as possible, just at the time that the observation pulse is applied. In the original version of Patt and Sykes [13] all the protons are first flipped 180° with a short pulse. This may be followed by a homogeneity-spoil pulse. Then the 45 to 90° (short) observation pulse is applied just at the time when the water magnetization has relaxed to zero on its way from inverted to normal. Since most macromolecules have T_1's much shorter than that of water, their spins will give almost a full signal at this time. This has recently been applied to millimolar nucleotide amino protons in ~ 100 % H_2O, and required an exceptionally uniform H_1 [14].

In my opinion this, and related methods [15], should be viewed as useful complements to the long pulse methods. The latter seem much easier to use once implemented, especially in pure H_2O. Furthermore, there is no disappearance of intensity distortion, by transfer of saturation, of resonances of protons which exchange rapidly with solvent.

We often use a closely related method [15], together with the long pulses, to reduce water spinning sidebands: we simply apply a weak presaturation pulse at the water frequency to partially saturate the water. This is used mainly for observation of resonances between 50 and 100 Hz from water, and, of course, not for rapidly exchanging protons whose resonances fortunately are usually further than this from water.

VI. Artifacts, Limitations, and Comparisons

It is of some interest to understand the limitations of different techniques, in view of the expense of setting up and running an NMR spectrometer. In this section I will present as much information as I can, based on one-sided experience, on this point.

a. Long pulse Techniques. Because the solvent signals are suppressed at the receiver system input by a factor of 100 or more during the entire free induction decay, this technique is *easy*, contrary to popular belief. The long low power pulse is the only absolutely essential element, but the other improvements which we made (quadrature detection, perfect gating, phase and amplitude correction, and digital timing) are useful from other points of view. It is easy for untrained operators to use the method since the pulse amplitude and length can be roughly preset and the operator can then fine tune for smallest water and protein free induction signal. In H_2O solvent, the baseline is not critically dependent on rf phase. It does vary appreciably if the pulse length varies by about 1 % or more, or if the difference between the solvent frequency and the transmitter frequency varies by more than 1 %.

As I write this, we are struggling with the first problem that has proven difficult: at 270 MHz there is a randomly varying H_2O signal after the pulse in our spectrometer. Apparently this is due to phase noise in our frequency synthesizer which is based on phase-locked loops. The single-sideband synthesis described in Ref. [3] may be more practical for this technique. High order spinning sidebands are also a more serious problem at 270 MHz.

The spectra in Fig. 5a, b and c are intended to show the power and versatility of the technique. Equivalent results are obtainable at 90 MHz. Fig. 5d shows the kind of artifact which occurs with lines which are not narrow compared to the spectral width. The negative wings on each side of the line are always comparable in area to the positive center line, integrated over the entire, spectral width, so that they are noticeable only for such borad lines. The same wing occurs downfield of the lysozyme aromatic and amide region in Fig. 5e. Such a spectrum would not yield a correct overall shape or integral for the aromatic region, but the sharp lines on top would be correctly reproduced.

The baseline artifact in Figs. 5d and e results from the impossibility of phase-correcting for the long pulse over the whole spectrum without rotating the dispersive wings of the line onto the real axis which we plot. It is impossible to correct for this effect completely; this would require exact knowledge of the NMR signal *during* the long pulse. We cannot detect this signal because of interference from the transmitter pulse, but even if we did, we would not get much better spectra because the rf phase setting would then become more critical. Long pulse spectra are relatively phase-independent even in H_2O, relative to swept spectra, because the signal after the pulse is always small.

b. Correlation Methods. For undemanding applications, in deuterated solvent, correlation spectroscopy is likely to be comparable to FTNMR in sensitivity, convenience, and ease of implementation. It appears to me that measurement of short T_1's may be more difficult to implement in correlation spectroscopy.

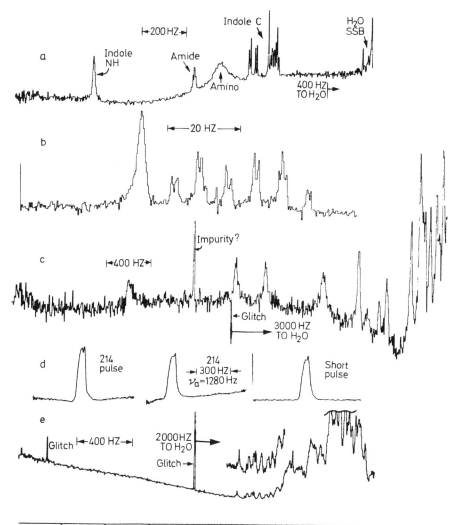

Fig. 5 a–e. Spectra obtained with the "214" long pulse (Ref. [11]) in 95% H_2O, 5% D_2O. No exponential or trapezoidal multiplication, artificial baseline tilt, delay, or artificial curvature used. (a) Tryptophanyl glycine, 10 mM, ph 1.1, entire downfield region, under conditions which enhance the amino resonance. 90 transients, one minute, 2560 Hz complex advance frequency, 1024 complex input points (2048 total input points). Resonances other than the amino resonance are decreased in amplitude because of insufficient recovery time. (b) A portion of the spectrum of the same sample under high-resolution conditions. 32 transients, two minutes, 1280 Hz advance frequency, 4096 input points. (c) Ferricytochrome c, 2 mM, pH ~7, a portion of the downfield region. 320 transients, one minute, 5120 Hz advance frequency, 1024 input points. The prominent lines in the center of the spectrum are all single protons near the heme. (d) The H_2O line, artificially broadened and run at 1 kHz spectral width. The right hand spectrum was run with a weak short pulse. The left and center spectra were run with a weak 214 pulse using two different f_1 settings, to show the baseline distortion. (e) Lysozyme, 4 mM, pH 3, a portion of the downfield region. 280 transients, one minute, 5120 Hz advance frequency, 1024 input points. The small peaks downfield of the aromatic region are tryptophon indole N protons; these are shown expanded vertically in the inset.

For samples dissolved in H_2O solvent, the H_2O signal going into the receiver is likely to be larger for correlation spectroscopy than in the long pulse method, though not as large as in short pulse FTNMR. As in swept NMR, this signal must be removed from the final spectrum by phase rotation of the spectrum. Thus the phase setting is more critical than in long-pulse FTNMR. Below saturation the method should yield perfect spectra, free of baseline curvature. I do not know whether it can yield spectra better than those obtainable from the long pulse methods in the same time, for samples with broad lines dissolved in H_2O.

VII. Acknowledgements

Most of the innovations described herein were developed in collaboration with Dr. Raj K. Gupta at the former IBM Watson Laboratory at Columbia University, and supported by that Laboratory. The spectrometer at Brandeis was built largely by Mrs. Sara Kunz and its development supported by grants from the National Science Foundation, the U.S. Public Health Service, and the Research Corporation.

Appendix A. Aspects of Conventional FTNMR

Convenience: Much commercial development has been directed toward making FT spectrometers more convenient and less formidable looking to the novice — and rightly so. An expensive and heavily used instrument can be used more efficiently if one does not have to fight it. Modern FTNMR spectrometers are now as easy to operate as swept instruments.

Phasing: Despite these developments it is still necessary for the operator to control a frequency-proportional phase rotation of the spectrum by manipulating the computer, on commercial instruments, even though a method for pre-computing this correction was presented in Ref. [3]. A strategy was also presented for pre-setting the frequency-independent phase shift. In the case of our present instrument this latter adjustment is often done by turning the phase knob while looking at a repetitive non-accumulating display of the real part of the FT of short blocks of transients. Once adjusted, this phase setting is stable and reproducible from day to day as long as the buffer is not changed.

On line display: It is very convenient and time-saving to be able to see the FT of an accumulated run. Our instrument now FT's every 16 transients and displays a running normalized spectrum from the frequency-domain accumulation. Presetting of the phase means that only the real part of the FT need be stored in core, saving core space. This "block averaging" method avoids memory overflow problems for strong lines. Double precision storage is also easier in the frequency domain.

Phase Reversal: In the Bruker WH90 the transmitter pulse phase is reversed every other event; this reverses the signal phase and is compensated for by the computer. This reduces or eliminates problems arising from poor gating or signal offset in the receiver system.

Gating: "Perfect" gating is desirable though lacking on many existing instruments. Any carrier signal leaking through the gate will give a glitch, or noise, on the output if it is coherently, or randomly, modulated. Ref. [2] gives strategies for perfect gating, which is also necessary for quadrature detection to avoid a large glitch at center band.

Lock: should be time-shared, with single sideband detection. Quality phase detectors, and dummy gating of the receiver at twice the transmitter pulse repetition rate [3], can yield a lock signal with no amplitude or phase-dependent baseline shift. If the lock gating is synchronized with the signal sampling rate, or a multiple thereof, then any possible leakage of lock transients into the proton signal channel will do no harm.

Spoil pulse: is produced by running a current pulse through a shim coil to momentarily spoil the field homogeniety. It is desirable to put such a pulse between the pre-irradiation and the observation pulse to avoid spin echoes, and also to dampen any signals stimulated by the preirradiation. This is especially useful when strong solvent lines are present.

Automation: Automation of both delays and double irradiation parameters in a series of runs is now commercially available. It is most efficient to build a completely separate control, timing, and automation circuit which is only loosely controlled by the computer. A versatile timer can be built using a single preset counter which modifies its own clock and preset value during different pulses [3].

Output. The advent of cheap microprocessors and shift register memories should make it worthwhile also to separate the time-consuming plotting, and perhaps display, functions from more demanding on-line computation by use of low-speed serial access output memories.

Appendix B. Phase and Amplitude Corrections in FTNMR

The filter correction is generated by grounding the imaginary filter input and connecting the real input to a μsec pulse generator. This pulse must go cleanly to zero and this is accomplished by running it through a diode which is reversed when the pulse is off. The pulse is most easily generated from the f_1 gate pulse. The output of the real filter is accumulated at moderate resolution (1 K) for many transients, as a free induction decay would be, but at a sampling rate ν_a which is several times the filter band width. Before FT, the imaginary signal stored in core is zeroed, while the baseline for the real signal is established by averaging the last hundred points of so, and is subtracted. Then all but the first one or two hundred points are zeroed in the real channel. The FT is performed, yielding the filter frequency response over a wide frequency range. The point at zero frequency is out of line because the baseline is not perfect, and is set equal to the average of the first few real frequency domain points. Then a computer program extracts a sample of twenty six points in the positive frequency response curve, takes the complex reciprocal of these, and multiplies all by the same constant to achieve a conveniently normalized function which, when interpolated, and used to multiply an uncorrected spectrum, would correct it. Negative frequency points are the complex conjugates of the corresponding positive frequency points. However, the differences between each of the successive points are what we actually stored. The programs which generate such a table are not stored the core after the table is complete.

The on-line phase correction starts correcting the spectrum at zero audio frequency, correcting points at equal positive and negative frequencies together. The output spectral points from zero to $-(1/2)\nu_a$ are derived from the complex FT points at a nominal frequency from ν_a down to $1/2\,\nu_a$ Pairwise phase correction saves time because the corresponding phase corrections are mutual complex conjugates. The phase correction function is resurrected from the difference table as the program proceeds. For each point of a 2 K, or smaller, spectrum a new phase correction function point is calculated by adding an increment to the previous one. Then this complex number multiplies the complex FT point for positive frequency, and its complex conjugate multiplies that for negative frequency. For 4 K (or 8 K) spectra a new phase correction point is calculated only every 2 (or 4) times. For spectra of less than 2 K in size the stored phase function differences must be renormalized by a number inversely proportional to the spectral size.

The long pulse correction is obtained for the corresponding twenty-six frequency points by calculating the position of a unit magnetization which is initially in the z direction, after precessing about the effective field for a time τ. The effective field is, of course, different for each frequency. A function of frequency is produced whose imaginary part is the final component of magnetization parallel to H_1 in the rotating frame, while its real part is the component perpendicular to H_1 and H_0. This function is normalized and its reciprocal taken. This reciprocal is multiplied by the reciprocal of the filter response to form a combined correction and then differences between the twenty-six successive frequency points are stored as before. A library of such corrections, for different values of τ, is stored in memory blocks of fifty words each. The on-line correction routine is the same as before, but simply uses a different table to resurrect the phase correction. The program which generates the tables (which requires floating point arithmetic) is not normally stored in core memory.

References

[1] Ernst, R. R., Anderson, W. A.: Rev. Sci. Instr. *37*, 93 (1966).
[2] Farrer, T. C., Becker, E. D.: Pulse and Fourier Transform NMR. New York: Academic Press 1971.
[3] Redfield, A. G., Gupta, R. K.: Advan. Magnet. Res. *5*, 81 (1971).
[4] Tomlinson, B. L., Hill, H. D. W.: J. Chem. Phys. *59*, 1775 (1973).
[5] Dadock, J., Sprecher, R. F.: J. Magnet. Res. *13*, 243 (1974).
Gupta, R. V., Ferretti, J. A., Becker, R. A.: J. Magnet. Res. *13*, 275 (1974).
[6] Wilson, D. M., Olson, R. W., Burlingame, A.: Rev. Sci. Instr. *45*, 1095 (1974).
[7] Stejskal, E. O., Schaefer, J.: J. Magnet. Res. *14*, 160 (1974).
[8] Redfield, A. G., Kunz, S. D.: J. Magnet. Res. *19*, 250 (1975).
[9] Allerhard, A. Childers, R. F., Oldfield, E.: J. Magnet. Res. *11*, 272 (1973).
[10] Alexander, S.: Rev. Sci. Instr. *32*, 1066 (1961).
[11] Grunwald, E., Ralph, E. K.: J. Am. Chem. Soc. *91*, 2422 (1969).
[12] Redfield, A. G., Kunz, S. D., Ralph, E. K.: J. Magn. Res. *19*, 114 (1975).
[13] Patt, S. L. O., Sykes, B. D.: J. Chem. Phys. *56*, 3182 (1971).
[14] Krugh, T. R.: Biochem. Biophys. Res. Commun. *62*, 1025 (1975).
[15] Campbell, I. D., Dobson, C. M., Genimet, G., Williams, R. J. P: FEBS Letters *49*, 115 (1974)
Bleich, H. E., Glasel, J. A.: J. Magnet. Res. *18*, 401 (1975).

Author Index Volumes 1—13

Armstrong, R. L.: Longitudinal Nuclear Spin Relaxation Time Measurements in Molecular Gases. 13, 71–95 (1976).

Bergmann, K.: Untersuchung von Beweglichkeiten in Polymeren durch NMR. 4, 233–246 (1971).

Bene, G. J.: Nuclear Spins and Non Resonant Electromagnetic Phenomena. 13, 45–54 (1976).

Blinc, R.: Spin-Lattice Relaxation in Nematic Liquid Crystals Via the Modulation of the Intramolecular Dipolar Interactions by Order Fluctuations. 13, 97–111 (1976)

Bovey, F. A.: High Resolution NMR Spectroscopy of Polymers. 4, 1–9 (1971).

Cantow, H.J., Elgert, K. F., Seiler, E., Friebolin, H.: NMR-Untersuchungen an Poly-α-Methylstyrol und dessen Copolymeren mit Butadien. 4, 21–46 (1971).

Clough, S.: NMR Studies of Molecular Tunnelling. 13, 113–123 (1976).

Connor, T. M.: Magnetic Relaxation in Polymers. The Rotating Frame Method. 4, 247–270 (1971).

Diehl, P., Kellerhals, H., Lustig, E.: Computer Assistance in the Analysis of High-Resolution NMR Spectra. 6, 1–96 (1972).

Diehl, P., Khetrapal, C. L.: NMR Studies of Molecules Oriented in the Nematic Phase of Liquid Crystals. 1, 1–96 (1969).

Fischer, H.: ESR-Untersuchungen an Hochpolymeren. 4, 301–309 (1971).

Forsén, S., Lindman, B.: Chlorine, Bromine and Iodine NMR. Physico-Chemical and Biological Applications. 12, 1–353 (1976).

Forslind, E.: Nuclear Magnetic Resonance Wide in Line Studies of Water Sorption and Hydrogen Bonding in Cellulose. 4, 145–166 (1971).

Guillot, J.: Penultimate Effects in Radical Copolymerization I – Kinetical Study. 4, 109–118 (1971).

Hahn, E. L.: Macroscopic Dipole Coherence Phenomena. 13, 31–44 (1976).

Harwood, H. J.: Problems of Aromatic Copolymer Structure. 4, 71–99 (1971).

Hilbers, C. W., MacLean, C.: NMR of Molecules Oriented in Electric Fields. 7, 1–52 (1972).

Hill, H. A. O.: The Proton Magnetic Resonance Spectroscopy of Proteins. 4, 167–179 (1971).

Hoffmann, R. A., Forsén, S., Gestblom, B.: Analysis of NMR Spectra. 5, 1–165 (1971).

Jeener, J.: Thermodynamics of Spin Systems in Solids. An Elementary Introduction. 13, 13–21 (1976).

Jones, R. G.: The Use of Symmetry in Nuclear Magnetic Resonance, 1, 97–174 (1969).

Kanert, O., Mehring, M.: Static Quadrupole Effects in Disordered Cubic Solids. 3, 1–81 (1971).

Keller, H. J.: NMR-Untersuchungen an Komplexverbindungen. 2, 1–88 (1970).

Khetrapal, Kunwar, Tracey, Diehl: Nuclear Magnetic Resonance Studies in Lyotropic Liquid Crystals. 9, 1–85 (1975).

Klesper, E., Gronski, W., Johnsen, A.: Complete Triad Assignment of Methylmethacrylate-Methacrylic Acid Copolymers. 4, 47–69 (1971).

Kosfeld, R., Mylius, U. v.: Linienbreiten- und Relaxationsphänomene bei der NMR-Festkörperspektroskopie. 4, 181–208 (1971).

McCourt, F. R.: Nuclear Spin Relaxation in Molecular Hydrogen. 13, 55–70. (1976).

Mehring, M.: High Resolution NMR Spectroscopy in Solids. 11, 1–243 (1976).

Noack, F.: Nuclear Magnetic Relaxation Spectroscopy. 3, 83–144 (1971).

Pfeifer, H.: Nuclear Magnetic Resonance and Relaxation of Molecules Adsorbed on Solids. 7, 53–153 (1972).

Pham, Q. T.: The Cotacticity of (Acrylonitrile-Methyl-Methacrylate) Copolymer by NMR Spectroscopy. 4, 119–128 (1971).

Pintar, M M.: Effect of Molecular Tunnelling on NMR Absorption and Relaxation. 13, 125–136 (1976).

Redfield, A. G.: A Guide to Relaxation Theory. 13, 1–12 (1976).

Redfield, A. G.: How to Build a Fourier Transform NMR Spectrometer for Biochemical Applications. 13, 137–152 (1976).

Richard, C., Granger, P.: Chemically Induced Dynamic Nuclear and Electron Polarizations-CIDNP and CIPED. 8, 1–127 (1974).

Rummens, F. H. A.: Van der Waals Forces in NMR Intermolecular Effects. 10, 1–118 (1975).

Shimanouchi, T.: Conformations of Polymer Chains as Revealed by Infrared Spectroscopy. 4, 287–299 (1971).

Slichter, W. P.: NMR Studies of Solid Polymers. 4, 209–231 (1971).

Tosi, C.: New Concepts in Copolymer Statistics. 4, 129–144 (1971).

Waugh, J. S.: Coherent Averaging and Double Resonance in Solids. 13, 23–30 (1976).

Williams, G., Watts, D. C.: Some Aspects of the Dielectric Relaxation of Amorphous Polymers Including the Effects of a Hydrostatic Pressure. 4, 271–285 (1971).

Zambelli, A.: Research of Homopolymers and Copolymers of Propylene. 4, 101–108 (1971).

NMR

**Basic Principles and Progress
Grundlagen und Fortschritte
Vols. 1–12**

Editors:
P. Diehl, E. Fluck, R. Kosfeld

Nuclear magnetic resonance spectroscopy, which has evolved only within the last 20 years, has become one of the very important tools in chemistry and physics. The literature on its theory and application has grown immensely and a comprehensive and adequate treatment of all branches by one author, or even by several, becomes increasingly difficult.
This series presents articles written by experts working in various fields of nuclear magnetic resonance spectroscopy, and contains review articles as well as progress reports and original work. Its main aim, however, is to fill a gap existing in the literature by publishing articles written by specialists, which take the reader from the introductory stage to the latest development in the field.

Volume 1
P. DIEHL, C. L. KHETRAPAL
NMR Studies of Molecules Oriented in the Nematic Phase of Liquid Crystals

R. G. JONES
The Use of Symmetry in Nuclear Magnetic Resonance
53 figures. V, 174 pages. 1969

Volume 2
H. J. KELLER
NMR-Untersuchungen an Komplexverbindungen
22 Abbildungen. III, 88 Seiten. 1970

Volume 3
O. KANERT, M. MEHRING
Static Quadrupole Effects in Disordered Cubic Solids

F. NOACK
Nuclear Magnetic Relaxation Spectroscopy
73 figures. V, 144 pages. 1971

Volume 4
Natural and Synthetic High Polymers
Lectures presented at the Seventh Colloquium on NMR Spectroscopy, April 13-17, 1970, Aachen
202 figures. X, 309 pages. 1971

Volume 5
R. A. HOFFMANN, S. FORSÉN, B. GESTBLOM
Analysis of NMR Spectra. A Guide for Chemists
63 figures. III, 165 pages. 1971

Volume 6
P. DIEHL, H. KELLERHALS, E. LUSTIG
Computer Assistance in the Analysis of High-Resolution NMR Spectra
11 figures. III, 96 pages. 1972

Volume 7
C. W. HILBERS, C. MACLEAN
NMR of Molecules Oriented in Electric Fields

H. PFEIFFER
Nuclear Magnetic Resonance and Relaxation of Molecules Adsorbed on Solids
56 figures. V, 153 pages. 1972

Volume 8
C. RICHARD, P. GRANGER
Chemically Induced Dynamic Nuclear and Electron Polarizations – CIDNP and CIDEP
26 figures. II, 127 pages. 1974

Volume 9
Lyotropic Liquid Crystals
18 figures. 3 tables
IV, 85 pages. 1975

Volume 10
Van der Waals Forces and Shielding Effects
13 figures. 46 tables. II, 118 pages. 1975

Volume 11
M. MEHRING
High Resolution NMR in Solids
104 figures. Approx. 240 pages. 1976

Volume 12
S. FORSÉN, B. LINDMAN
Chlorine, Bromine and Iodine NMR. Physico-Chemical and Biological Applications
72 figures. 42 tables.
Approx. 400 pages. 1976

Springer-Verlag
Berlin
Heidelberg
New York